WHY?

WHY?

Answers to everyday
scientific questions

JOEL LEVY

Michael O'Mara Books Limited

First published in Great Britain in 2012 by
Michael O'Mara Books Limited
9 Lion Yard
Tremadoc Road
London SW4 7NQ

A CIP catalogue record for this book is available from the
British Library.

Papers used by Michael O'Mara Books Limited are natural,
recyclable products made from wood grown in sustainable
forests. The manufacturing processes conform to the
environmental regulations of the country of origin.

ISBN: 978-1-84317-951-1 in hardback print format
ISBN: 978-1-84317-977-1 in EPub format
ISBN: 978-1-84317-978-8 in Mobipocket format

1 3 5 7 9 10 8 6 4 2

Designed and typeset by Greg Stevenson
Illustrations by Greg Stevenson

Printed and bound in Great Britain by Clays Ltd, St Ives plc

www.mombooks.com

Contents

Introduction

'A generous and elevated mind is distinguished
by nothing more certainly than an eminent
degree of curiosity.' *Samuel Johnson*

'Be not curious in unnecessary matters,' the Bible tells us, 'for more things are showed unto thee than men understand.' Respectfully I disagree, and the fact that you've picked up this book suggests that your generous and elevated mind is also distinguished by an eminent degree of curiosity. Hopefully the answers in these pages will reward that curiosity. Did you know, for instance, that you get heavier when you sunbathe but when water freezes into ice it gets lighter; lobsters never get old; the average man is stronger than 99.9 per cent of women; the Earth is smoother than a billiard ball; and that contrary to popular belief giraffes are very chatty (but talk in tones too low for us to hear)?

These and many other bits of trivia – from the number of volcanoes erupting as you read this line (around 20) to the colour of the sky on Mars (red) – are revealed in the process of answering over fifty deceptively simple questions about nature.

Obviously there are an infinite number of questions that could have been asked, so why choose this particular sample? In selecting the questions I have tried to cover both 'classic' questions about nature – why is the sky blue? – and some that might not have previously occurred to you – why do we forget? I've also tried to cover a wide range of subject areas, from geology to psychology, cosmology to chemistry, with questions ranging from the mundane – why does my mobile lose signal? – to the esoteric – why does time go forward?

In answering the questions I have tried to provide several levels or layers of explanation. Each answer begins with (an attempt at) a one-line explanation, before giving more detail and, in some instances, challenging the basic premise of the question. For instance, did you know that there is no dark side of the moon? Then the answer is given in more detail, with a fuller explanation, and where possible some illuminating trivia.

An important distinction to bear in mind when reading the answers is the difference between proximate and ultimate explanations. A proximate explanation describes the immediate, direct cause for something. For instance, ice floats because it is less dense than liquid water. But proximate explanations often beg the question – i.e. why is ice less dense than water? This is where the ultimate explanation comes in, which details the root cause, in this example describing how the unique ability of water molecules to form a special type of bond with each other leads to the unusual properties of ice.

The search for ultimate explanations can steer us in exciting and unexpected directions, and you will see that some of the least promising questions turn out to have the most interesting answers. For example, why are women shorter than men? At first glance this doesn't look like the most interesting question in the book, but the search for an answer leads into strange territory, taking us on a journey through the sexual habits of apes, the logic of genetic inheritance and the importance of a father's involvement in raising girls.

Not all of the questions can be answered. Nobody really knows, for instance, why we dream or even the reason for sleep. Explanations for the expansion of the universe are theories – informed speculation on a cosmic scale. The nature and causes of global warming are extremely contentious, and in a book of bite-sized answers there obviously is not the space to do full justice to the range of opinion or evidence involved. Karl Popper pointed out that 'Our knowledge can only be finite, while our ignorance must necessarily be infinite,' but I would prefer to leave you with some consoling words of James Thurber: 'It is better to ask some of the questions than to know all of the answers.'

Why does the sun shine?

The sun shines because it is constantly exploding like a giant nuclear bomb.

A nuclear bomb explodes because of a process called nuclear fusion, in which the nuclei (the bits in the middle) of atoms are fused together, releasing huge amounts of energy. Some of this energy is in the form of heat, and some in the form of light. The sun is like a massive nuclear bomb exploding all the time, and so it gives out loads of heat and light – so much that it warms and brightens the Earth 92 million miles away, where we see and feel the sun's energy as sunshine.

The sun is a huge ball of gas, 2,000 trillion trillion tonnes of it (that's 333,000 times heavier than the Earth). Most of this gas is hydrogen. The enormous gravity of the sun squeezes this gas so tightly that the nuclei of the hydrogen atoms fuse together to form helium nuclei. In the process a tiny fraction of the mass of each nucleus is converted into energy, so that the sun is exploding more fiercely than 4 million nuclear bombs a second!

Burning away

The sun burns through so much fuel that it loses the equivalent of a supertanker every heartbeat. Fortunately the sun is so immense that this missing chunk of matter makes little difference. Even though it has been burning matter at this rate for billions of years, the sun has only lost 0.1 per cent of its mass, and it will be another 5 billion years before it runs out of hydrogen for fusion.

What exactly happens in the sun during nuclear fusion? The nucleus is the bit in the centre of an atom, which is normally surrounded by electrons (see page 141). The intense heat of the sun strips these electrons off atoms of hydrogen, leaving behind a big soup of hydrogen nuclei. Hydrogen is the smallest and simplest element, and its nuclei consist of single nuclear particles. Under the intense gravity of the sun, four of these nuclear particles are crushed together until they fuse into a new type of atomic nucleus – a helium nucleus. During this process, the sun generates energy at the same rate as 400 million billion power stations working at full capacity (a billion is one thousand million).

Only in the centre of the sun is the gravitational pressure strong enough to trigger fusion. The sunshine we see and feel is generated by the intensely hot surface. If you heat a metal spoon to a high enough temperature it will become white hot and glow with intense heat and light; similarly, solar fusion is not the direct source of sunshine, but it is the ultimate cause.

If the sun is exploding with the force of more than 4 million nuclear bombs a second, why doesn't it fly apart? The answer is gravity. The size of the sun is the result of the balance between the outwards and inwards forces affecting it: gravity is pulling its matter in towards the centre, while the explosive power of nuclear fusion is blasting it outwards. When the sun does finally begin to run out of fuel, this balance will change and gravity will eventually triumph; the sun will end up as a white dwarf – a solidly compacted mass of atomic nuclei, gradually cooling down until it is a dead cinder floating in space.

Why are plants green?

Plants are green because they soak up red and blue light. If you take the red and blue colours out of white light, what you are left with is green.

Humans and other animals get their energy by eating and digesting food, but plants make their own food by harvesting sunlight. They use the energy from sunlight to drive a chemical reaction called photosynthesis that converts carbon dioxide and water into sugar.

Sunlight is a mixture of different colours – or wavelengths – that combine to form white light. When white sunlight falls on a leaf, the red and blue wavelengths are absorbed by a pigment called chlorophyll. A pigment is a chemical that is very good at absorbing some wavelengths of light and reflecting others, giving it an intense colour. Once the pigment has done its work, green light is all that is left to reflect back to our eyes.

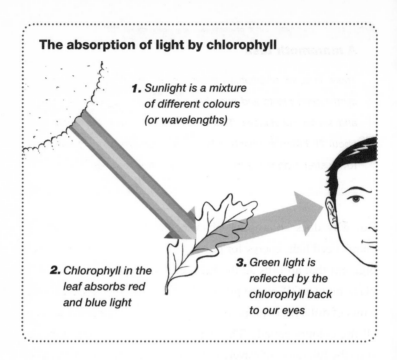

The absorption of light by chlorophyll

1. Sunlight is a mixture of different colours (or wavelengths)

2. Chlorophyll in the leaf absorbs red and blue light

3. Green light is reflected by the chlorophyll back to our eyes

Chlorophyll is a complex molecule that has evolved to capture light energy and transfer it to other molecules. It works in a similar fashion to a dish you might use to pick up television signals. A satellite dish collects radio waves carrying television transmissions from a wide area and focuses them onto a single point, concentrating lots of weak signals into a strong one. Similarly, chlorophyll collects light over a wide area and concentrates the energy into a single channel. This concentrated light energy is powerful enough to drive a biochemical cycle that makes sugars, which in turn act as a kind of biochemical fuel store, easily burned by the plant's cells to release energy when it is needed.

A mammoth task

There is a lot of photosynthesis going on; every year photosynthesizing plants and microbes make 105 petagrams of sugar and similar substances. That's 105 billion tonnes, equivalent to about 26 billion elephants, which is enough elephants to make a tower stretching to the moon and back 136 times over.

But why does chlorophyll absorb only red and blue light? If plants need light energy for photosynthesis, why don't they use a pigment that absorbs all the light; in other words, why not use a black pigment? Why are plants green and not black? Sunshine is a mix of different colours (or wavelengths), but it doesn't contain all the colours equally. The spectrum of light coming from the sun (the full range of wavelengths) is more intense in some bands than in others. A pigment that absorbs across all bands equally would be inefficient. It takes energy to build pigments, and the more colours the pigment has to absorb, the more energy it costs to build. Evolution favours the most efficient solution to a problem, and plants that wasted energy-building pigments to absorb weak colours would be out-competed by more efficient plants. In theory, the colours absorbed most strongly by chlorophyll are those that are most intense in sunlight. An obvious explanation for why chlorophyll reflects green and absorbs red and blue would be that red and blue are the most intense colours in the spectrum of sunshine and green is the weakest.

In reality, the opposite is true. The most intense part of the solar spectrum is the green-yellow band. Why are most plants not using a pigment that absorbs it? No one really knows, but there are various explanations. It is possible that the green part of sunshine is too intense, and the delicate molecules involved in photosynthesis might be damaged by it (by analogy, humans need some oxygen in the air that we breathe, but oxygen is a very reactive element and becomes toxic at high levels. If you are given 100 per cent oxygen to breathe you will soon die). Another explanation is that when chlorophyll originally evolved, the Earth was already dominated by micro-organisms using a different pigment, called retinol, which does absorb green-yellow light, so that the dominant colour of life on the planet was purple. These purple microbes soaked up all the green-yellow light, leaving only red and blue light for chlorophyll-bearing microbes to exploit.

Why are there 365 days in a year?

There are 365 days in a year because the Earth turns round 365 times in the time it takes to go all the way around the sun.

To be precise, there are 365.25 days in a year according to NASA but our calendars only make room for whole days, so the extra quarter from each year is saved up and used all at once every four years in a leap year, when the calendar has 366 days. The actual time it takes for the Earth to go around the sun and get back to the exact same spot is 1.0000174 years.

Other planets have longer or shorter years, depending on how far out from the sun they are. Neptune is 4.5 billion km (2.8 billion miles) from the sun and takes 164.79 Earth years to go around it, although its day is shorter than an Earth day (just sixteen hours long) because it spins round faster. Mercury takes just 88 days to complete an orbit at a speed of 173,326 kph (107,700 mph) as it is the closest planet to the sun and experiences much greater gravitational pull. However, Earth is no slouch: in a year it travels 945.5 million km (587.5 million miles) around the sun at about 107, 826 kph (67,000 mph).

The Earth didn't always take 365.25 days to orbit the sun because early in its history it spun round much faster. This meant that days were shorter, so there were more of them in a year. Then a few million years after it formed, the young Earth was hit by an object as big as a small planet causing huge amounts of rock to fly out into space. This rock soon collected together to form the moon, and ever since the moon's gravity has been slowing down the rotation of the Earth. Before the moon was formed, an Earth day was just six hours long. If the moon had never existed then today would be just eight hours long, and there would be 1,095 days in a year (see pages 87-8).

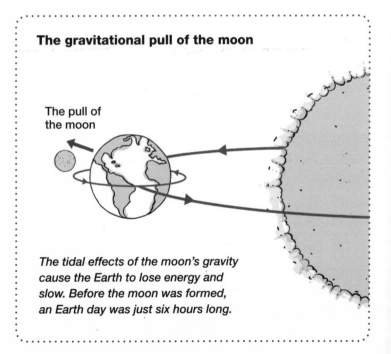

The gravitational pull of the moon

The pull of
the moon

The tidal effects of the moon's gravity cause the Earth to lose energy and slow. Before the moon was formed, an Earth day was just six hours long.

At least a day is twenty-four hours long, right? Not really. Firstly, the rotation of the Earth is still slowing down by about half a second per century. Secondly, the time it takes the Earth to spin right round on its axis is actually twenty-three hours, fifty-six minutes and four seconds. However, during this time the Earth has moved in space, so to spin round to the point where the sun is back in the same place in the sky takes another three minutes and fifty-six seconds, adding up to twenty-four hours.

Why does water freeze?

Water freezes because water molecules stick to one another when they get cold and slow down.

Water is strange stuff; in fact, it is one of the strangest substances in the known universe. Its odd properties are essential for the evolution and survival of life on Earth, particularly its ability to form a weak connection called a hydrogen or H-bond. A water molecule is made up of an oxygen atom connected to two hydrogen atoms, and the way that these atoms share their electrons turns the water molecule into a sort of mini-magnet. Just as two magnets will stick to each other, so a water molecule will stick to other water molecules with H-bonds.

A sticky business

In liquid form, the individual water molecules are quite energetic, forming and breaking hydrogen bonds with one another at great speed. In other words, they are sticky enough to adhere as a liquid (not flying off as in a gas) but not sticky enough to stay put. When water cools, its molecules lose energy and slow down, and their H-bonds stay stuck together.

Soon, each water molecule is attached to four others in a rigid lattice or network, forming an ice crystal. The lattice is pretty chaotic, however. An average ice cube contains about 60 billion trillion molecules. If you had spent every week since the big bang coming up with a different arrangement of these molecules you would not have exhausted the number of possible arrangements. Every single ice cube ever created has probably had a different arrangement of water molecules.

Almost any substance will 'freeze' if cooled down enough. But why does water freeze so easily? If it weren't for the H-bonds, water would boil at -90°C (-130°F), and there would be almost no liquid water – and no life – on Earth. An even more intriguing question is, 'why does water expand when it freezes into ice, whereas most substances shrink as they cool down?' See page 115 to find out why.

Why do tides ebb and flow?

Tides ebb and flow because the tidal pull of the moon stays in one place while the Earth spins round.

Imagine you are riding on a merry-go-round. It spins on a patch of ground that is flat aside from a big bump. Every time your part of the merry-go-round passes over the bump you rise up in the air. Now imagine that the merry-go-round is walled in. You can't see out and you don't realize that you are actually going round (everything you can see is going round at the same speed, so to you it seems as though you are not moving). All you know is that every 30 seconds the ground rises up. This is similar to someone sitting at the beach watching the ebb and flow of the tide. It seems like the water is moving while the beach is stationary, but in fact the beach is moving along with the rest of the Earth's surface, and it is the 'bump' that is stationary. In this instance, the 'bump' is the tidal pull of the moon.

Tides are caused by the gravitational attraction between the moon and the Earth. The ground is too solid to move much in response to the moon's tidal pull, but the oceans are liquid and they get pulled towards it. The result is that the oceans bulge out towards the moon. The moon does go around the Earth, but it moves pretty slowly compared to the daily rotation of the Earth, so for all intents and purposes the moon stays in the same place as the Earth spins round. This means that the tidal bulge caused by the moon also stays in the same place. That beach you are sitting on passes through the bulge once every twenty-four hours, and there is also a complementary bulge on the other side of the planet from the moon, so your beach actually passes through two bulges every twenty-four hours, which is why there are two tides a day. As your beach passes into the bulge the sea level starts to rise – this is the tide coming in or flowing – until it reaches a maximum height: high tide. Then as the beach passes out of the bulge the tide goes out or ebbs.

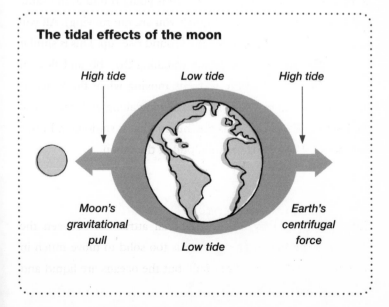

The tidal effects of the moon

High tide   Low tide   High tide

Moon's gravitational pull

Earth's centrifugal force

Low tide

Low tide

Tides are also affected by the gravitational pull of the sun, and by the depth of the ocean and the shape of the land. At the Bay of Fundy in Nova Scotia the shape of the coastline and sea bottom combine to create the world's biggest tidal range (difference between high and low tide), up to 17 m (18.6 yd) – this is the height of a five-storey building.

Why do things burn?

Things burn because by combining with oxygen they can move to a lower energy state.

Burning, technically known as combustion, is a chemical reaction where a substance combines with oxygen and releases energy in the form of heat and light, which we see as flames. The substance starts off in a higher energy state, and by combining with oxygen ends up in a lower energy state. In terms of its energy state, matter is like water – it flows downhill. Wherever possible, substances move from a higher to a lower and more stable energy state.

The simplest example of a combustion reaction is when hydrogen burns in oxygen. The two gases combine to produce water, and water is much more stable than either oxygen or hydrogen: try setting fire to it and see for yourself. By contrast, both hydrogen and oxygen are highly inflammable gases because they exist at much higher energy states than water. The difference between the before and after energy states is the amount of energy that is released during the combustion reaction. A lot of energy is released when hydrogen burns in oxygen and this makes it a good rocket fuel. The Saturn V rockets that blasted the Apollo moon-landing missions into space used hydrogen and oxygen as fuel – over 3.6 million litres (960,000 gallons) of it! The first stage tank alone contained enough liquid oxygen to fill fifty-four railroad tank cars.

Catching fire

If burning moves a substance from a higher to a lower energy state, why doesn't combustible stuff like wood burst into flames spontaneously? In fact, it is quite hard to set fire to wood, as anyone who has tried to get a campfire going in the rain will tell you. Some substances will catch fire spontaneously, but they tend not to hang around too long. Most require a burst of energy to get the combustion reaction going. This is called activation energy.

The changing energy states of burning wood are like water moving from the top of a slope to the bottom; activation energy is the hump that water has to get over before it can run downhill. Once it gets a little boost to get it over the hump, it will carry on running downhill until it reaches flat ground. Similarly, once the wood gets a boost from a lit match, it will burn to ash. In energy terms, ash is the equivalent of flat ground. In fact, with wood the picture is more complex, because wood itself doesn't burn directly; instead once it reaches ignition temperature of about 260°C (500°F) the heat causes it to break up and some of it turns into a gas, and it is this gas that burns.

Why does E=mc²?

E=mc² because mass and energy are two sides of the same coin.

Einstein's famous equation is a mathematical way to express something called the equivalence of mass and energy. Although in everyday life we think of mass and energy as being very different things, physicists consider them to be different properties of the same thing.

In Einstein's equation, the E stands for energy, m stands for mass and c stands for the speed of light. His equation says that mass and energy are interchangeable, and if you want to figure out how much energy is equivalent to a given amount of mass (or vice versa), you do the conversion by multiplying or dividing by the speed of light (c) squared. Often the speed of light is measured in miles per second, so that $c^2 = 186,282 \times 186,282 = 34,700,983,524$ m/s, which is a lot. But you can set the units in the equation to be what you want. If you set them so that $c=1$, then $E=mc^2$ becomes $E=m1^2$. The square of 1 is 1, so this is just the same as saying $E=m$ or energy = mass.

Einstein worked out his equation after he realized that an object becomes heavier when it gains energy (technically, it becomes more massive because weight and mass are not the same thing, but for everyday purposes we can say 'heavier'). When you sunbathe and heat up, you get heavier. When a cube of water freezes into ice, it gets lighter. But the amounts involved are so minute that you can't detect them. To work out how much mass you get from a given amount of energy, you rearrange Einstein's equation. Instead of $E=mc^2$, write it as $M=e/c^2$. In other words you have to divide energy by the square of 'c', the speed of light, which is a huge quantity, so you get a tiny quantity as the result.

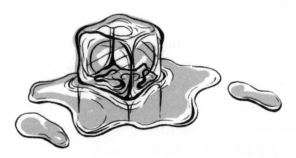

When you sunbathe and heat up, you get heavier. When a cube of water freezes into ice, it gets lighter.

For instance, a charged-up battery weighs about 0.0000000001 grams more than a battery that has been discharged. If you heat a gold bar weighing a kilogram up by 10°C (50°F) it gains about 0.000000000014 g. There are other types of energy apart from heat and chemical energy. To make something move, you give it kinetic energy. Every time you throw a ball, the ball gains kinetic energy and gets a tiny bit heavier. When a pitcher throws a baseball at 160 kph (100 mph) the ball gets 0.000000000002 grams heavier. The fact that things get heavier as they get faster is very important to the question of why nothing can travel faster than light – see page 146.

This relationship also works the other way round: mass is equivalent to energy, so when something gets less massive, it releases energy. We can say that the mass has been converted to energy. And because you can work out how much energy by multiplying the mass by the speed of light squared, you can see that even a tiny amount of mass can be converted into a lot of energy. This is the basis of the nuclear bomb, and also of the nuclear fusion reaction that powers the sun (see page 11). The energy content of just one kilogram of matter would be enough to lift the population of the Earth into space. If you were converted entirely into energy, you would explode more powerfully than thirty nuclear bombs.

Why do some things float?

An object floats if it weighs less than the water that occupies the same amount of space as it.

When a thing is less dense than water, it is buoyant. The 'thing' refers to a whole object – so a ship floats even though it is made of iron, which is denser than water. The overall density of the ship is reduced because the iron encloses a lot of air, which has a low density. The *Titanic* weighed 46,000 tons, but it floated because the water it displaced weighed even more. If the *Titanic* had been loaded to the funnels with lead, it would have weighed more than the water it was displacing and it would have sunk. Indeed, when an iceberg made a hole in the side and the *Titanic* filled up with water, it was no longer buoyant and it sank.

The least buoyant material in the universe is the stuff that makes up a neutron star. It is the densest material known, and indeed the densest material that can exist. A piece of neutron star the size of a sugar cube weighs more than the entire human race. If our sun was as dense as a neutron star, it would be crammed into a space smaller than Mount Everest.

Take the weight off

To experience greater buoyancy, float upon the Dead Sea, a salt lake on the border between Jordan, Israel and the West Bank. Because six times more salt is dissolved in it than ordinary seawater, the water in the Dead Sea is more dense, which makes you more buoyant, which in turn is why it is easier to float in it.

Legend has it that the principle of buoyancy was first recognized by the ancient Greek mathematician Archimedes, prompting his famous 'Eureka!' incident. The story goes that the king of Syracuse gave a lump of pure gold to a smith to make a laurel wreath, but suspected the man of cheating by stealing some of the gold and replacing it with silver. The king asked Archimedes to help unmask the fraud. Archimedes knew that a gold-silver alloy was less dense than pure gold, so if the wreath were not pure it would have to be bigger (i.e. have a greater volume) than a lump of gold that weighed the same. Unfortunately there was no way to calculate the volume of a complex shape such as a wreath. Archimedes pondered the conundrum while relaxing in the bath, and noticed that the more he reclined, the more water spilled out of the bath. He realized that the volume of water that spilled from the bath was an exact measure of the volume of space he took up, and that here was a principle that could be used to determine the volume of the wreath. He was so excited that he leapt naked from the bath with a cry of 'Eureka' (Greek for 'I've got it!') and capered down the street.

In the legend, Archimedes puts a lump of pure gold weighing the same as the wreath into a bowl filled to the brim with water. He then removes the gold and submerses the wreath. The bowl overflows, proving that the wreath has a greater volume and therefore lower density than the pure gold, and must be adulterated. The guilty goldsmith gets his comeuppance.

Commentators since Galileo have pointed out that the difference in volume between the wreath and the lump of gold would be so small that this method would not work in practice, but it is known from Archimedes' surviving writings that he did explore the principle of buoyancy, and so could plausibly have arrived at a cunning, water-displacement-based solution to the puzzle of the wonky wreath.

Why do apples fall down?

Apples fall because they are pulled by gravity towards the centre of the Earth.

Isaac Newton famously conceived of the force of gravity when struck on the head by a falling apple, or so the legend goes. Newton wasn't the first person to think about gravity; everybody knew that things fell down when you dropped them. His breakthrough was to realize that it was a fundamental force of nature, affecting everything from the fall of an apple to the orbit of the moon around the Earth. In the summer of 1666, Newton was sitting in his garden when he saw an apple fall from a tree and hit the ground. At this time he was thinking deeply about what kept the moon in its orbit, and it suddenly occurred to him that there must be a relationship between the force of gravity pulling the apple to Earth and the same force pulling the moon towards the Earth. He was even able to work out just how strong that force must be, and how it got weaker with distance.

The answer that Newton came up with is that gravity is a force of attraction between any and all objects, and that its strength is determined by the size of the objects and the distance between them. There is a force of gravity between an apple and another apple, but because they are so small it will be immeasurably minute. The Earth is pretty big, so for things that are close to it, like apples, the force of gravity between them is quite strong – strong enough to make an apple that comes loose from its tree accelerate towards the ground at a rate of about 10 m (10.94 yd) per second per second (i.e. every second that the apple falls, its speed increases by 10 m/s). Bigger planets have stronger gravity; on Jupiter, an apple would accelerate towards the ground at 26 m/s^2 (28.43 yd/s^2), and on the surface of the sun an apple would fall at 274 m/s^2 (299.65 yd/s^2), which means that it would take about a third of a second to be travelling as fast as a Formula 1 racing car.

About 240 years after Newton, Albert Einstein came along and changed our view of gravity. He said that the universe was made out of space-time, and that objects cause bulges in the fabric of space-time like balls sitting on a rubber sheet. A big ball, like the Earth, causes a very deep bulge, so that a smaller ball, like an apple, will roll down the slope into the bulge. The steepness of the bulge is what decides how quickly other balls will fall down it – i.e. how strong the force of gravity is. So Einstein might have said that the reason an apple falls to the ground is that it is following the contours of space-time, which are deformed by the mass of the Earth.

Why are there so many planets in our solar system?

There are so many planets in our solar system because our sun didn't grow too big, too fast, and this allowed space and time for a large number of planets to form.

There are eight planets in our solar system. There used to be nine but in 2006 the International Astronomical Union (IAU) decided that Pluto is too small to be considered a planet and should be reclassified as a dwarf planet or Kuiper Belt object. The Kuiper Belt is a zone of large objects in the outer reaches of the solar system, many of which are the same size or bigger than Pluto. The IAU concluded that if Pluto were to be classified as a planet, so should these other objects, and our solar system could be said to have a dozen or more planets.

The planets probably formed from something called an accretion disc. This started life billions of years ago as a giant cloud of gas and dust floating in space. Something disturbed the cloud – possibly shockwaves from a nearby exploding star – pushing some of the gas and dust closer together. When some of the cloud concentrated in one spot, its gravitational pull became stronger than the rest of it, and this concentrated area was quickly pulled into a big ball.

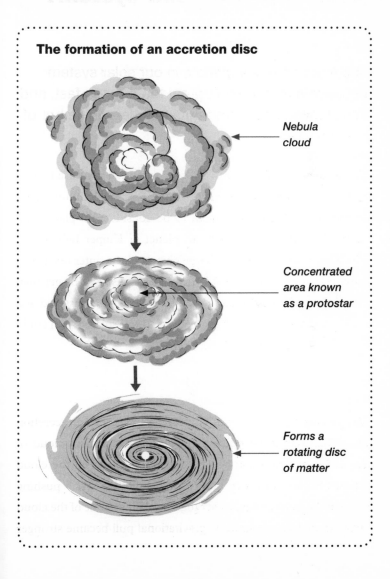

The formation of an accretion disc

Nebula cloud

Concentrated area known as a protostar

Forms a rotating disc of matter

As it got squished together this ball began to heat up, becoming a protostar – a young sun. What was left of the cloud got squeezed into a disc that rotated around the protostar, and this disc cooled down enough for the gas in it to freeze into ice and rock particles. The accretion disc gets its name from the way some of the larger particles in it start to accrete (i.e. pick up) other, smaller particles. They grow into large, asteroid-like objects, called planetesimals, and these in turn collide with each other; in our solar system, this process ended up with the eight planets we know today.

If the protostar had been bigger, it might have split into a double star and the two stars would probably have 'cleaned up' a lot of the accretion disc, leaving less material for planets. If the protostar had been hotter, the accretion disc would never have cooled down enough for gas to solidify into ice and rock and the planets might not have formed. There are probably lots of other things that could have turned out differently and would have changed the number of planets. For instance, if the orbit of Jupiter had been very slightly different it would have knocked most of the other planets out of orbit and sent them hurtling into space.

Why is the universe expanding?

The universe is expanding because it is filled with more dark energy than matter.

According to the big bang theory, the universe started off as an infinitely small point and expanded incredibly quickly in a process called inflation. To go from essentially nothing to about the size of a Space Hopper took just 10^{-34} seconds (that's 0.00 0000000000000000000000000000001 seconds!). This early universe expanded at 3 trillion trillion trillion times the speed of light. The original cause for the expansion of the universe was inflationary pressure from the big bang, thanks to which the universe is now about 150 billion light years across, even though it is only 13.7 billion years old.

As a result of the big bang, lots of matter appeared in the form of gas, dust, stars, galaxies, black holes, and so on. Matter causes gravity, and gravity is a force which attracts things together, so if the universe were made only of matter, then gravity would be acting against the outward pressure of big bang inflation. Astronomers would expect to see the expansion of the universe slowing down, or even going into reverse, but in fact they see the expansion accelerating. This strongly suggests that the universe must be stuffed full of some sort of energy, but the problem is that no one can see this energy and no one knows what it is.

In the dark

Scientists call this energy dark energy. They know that about 70 per cent of the universe must be made of dark energy, and another 25 per cent of a form of invisible matter called dark matter. All the visible matter in the universe, including all the galaxies, stars, planets, black holes and nebulae, adds up to just 5 per cent of the total. The dark energy is somehow causing the universe not just to expand, but to expand increasingly fast.

How do astronomers know how much stuff forms the universe? They measure something called the cosmic background microwave radiation (CBMR), which is a remnant of the big bang, like a very faint echo. The CBMR accounts for 99 per cent of the radiation in the universe. You can see it for yourself by switching on your television set but leaving it untuned – about 1 per cent of the static you can see is your television aerial picking up the CBMR.

Why can't we hear dog whistles?

We can't hear dog whistles because the sound they make is too high-pitched for human hearing.

Humans can hear a limited range of pitch or frequencies. Sound that is too high-pitched or too high-frequency for us to hear is known as ultrasound whereas sound that is too low is called infrasound. The pitch or frequency of a sound is measured in hertz (Hz) or vibrations per second; 1 Hz = one vibration per second. One thousand vibrations per second is called a kiloHertz, or kHz. Most people hear sounds within the range 20 Hz–20 kHz, although younger people have a wider range of hearing than older people, and women have a wider range than men. Dogs can hear much higher pitches than us, so they can hear dog whistles, which make sounds in the 20–22 kHz range. Dog whistles let dog trainers and owners make a loud noise that the dog can hear from a distance but which won't bother other humans. The whistle itself has no special power to command the dog; an animal will only respond if it's been trained to do so.

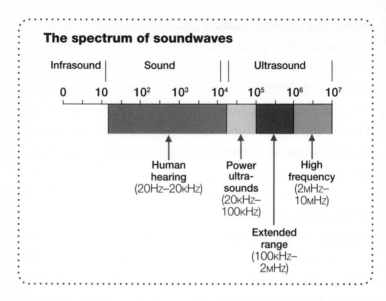

The spectrum of soundwaves

Infrasound | Sound | | Ultrasound |

0   10 10^2 10^3 10^4 10^5 10^6 10^7

Human
hearing
(20Hz–20kHz)

Power
ultra-
sounds
(20kHz–
100kHz)

High
frequency
(2mHz–
10mHz)

Extended
range
(100kHz–
2mHz)

Other animals that can hear ultrasound include bats and dolphins, which use ultrasonic clicks and whistles for echolocation, the animal version of sonar. By sending out sounds and measuring how long they take to bounce back, these animals can 'see' with sound. There are lots of animals that use infrasound, too. Giraffes are usually thought of as silent creatures, but in fact they are constantly chatting using infrasound. Many other large animals use infrasound; tigers use infrasound to stun their prey, and the infrasonic booms of elephants can be heard by other elephants over six miles away. Blue whales make such loud infrasonic noises that a human swimming next to the whale's throat would have his brain turned to jelly! Natural forces like earthquakes and tsunamis generate infrasound, and it is thought that animals are able to avoid being swept away by tsunamis because they pick up infrasonic warnings and flee before the wave hits.

So why can't humans hear infra- and ultrasounds? The answer is probably linked to the size of the human body and what we use sound for. Big animals like elephants and tigers need a lot of food, so they can't afford to have too many individuals living close together or they would quickly run out of things to eat. But they still need to communicate to find mates and warn off enemies, so they use infrasound because it can be heard a long way away. Being big makes it easier for them to generate loud low-frequency noise. For small animals it is easier to make high-pitched noises, which suit their needs better as they are useful for communicating across short distances. In terms of size, humans are in between, and we need to hear noises such as other people's voices, so the most useful frequency range for us is about 20 Hz–20 kHz. Evolution favours individuals who make the most efficient use of their resources, and growing a hearing system that can hear both infra- and ultrasound would be very expensive in terms of material and energy. The most efficient solution is to grow ears that only hear the most useful range of frequencies, and this is exactly what has happened.

Why does iron rust?

Iron rusts because iron combines with oxygen when water is around, forming a new substance called iron oxide, or rust.

Iron is easily converted into iron oxide for the same reason that wood burns: by combining with oxygen it achieves a lower and more stable energy state (see page 25). In fact, rusting can be described as a form of combustion: a very slow form of burning.

Another way of looking at rusting is as an electrolysis reaction, like the ones that take place in batteries. An electrolysis reaction is an electrochemical process – involving electricity and chemicals. Electricity is the movement of electrons (subatomic particles with negative charges) and in electrolysis the electrons come from a piece of metal called an anode. The anode combines with oxygen and releases an electron which travels through a fluid called an electrolyte. In rusting, the anode is the piece of iron and the electrolyte is water. If the water contains salt, it acts as an even better electrolyte, which is why salt speeds up rusting, which in turn is why things rust quicker when exposed to seawater. But salt isn't necessary – even pure water will dissolve carbon dioxide from the air to create a weak acid, and this acts as the electrolyte.

Lady Liberty

Other metals also form oxides, but they are not called rust. Copper forms a layer of oxide called a patina. One important difference between patina and rust is that patina protects the metal underneath. The skin of the Statue of Liberty is made of copper, and the patina which formed on it has protected it from corrosion so effectively that when the Statue was restored after a hundred years of exposure to rain and salt spray, the only bit that didn't need replacing was the skin. In all that time, the patina had grown to just 0.127 cm (0.05 in) thickness. Rust, on the other hand, does not protect underlying iron, so a piece of iron will eventually turn completely into iron oxide and crumble away entirely.

Steel is iron that has been strengthened by adding tiny amounts of carbon, but like iron it will rust. Stainless steel is a rust-proof steel alloy (a mixture), made by adding nickel and chromium, which bind to the iron atoms so that they can't bind with oxygen. Another way to protect steel is by covering it with a rust-proof coating. The Forth Railway Bridge, one of the world's largest steel structures, has 400,000 m² (478,396 yd²) of surface that has to be protected from corrosion, so it is 'painted' with a layer of zinc and another of glass flakes. Without this protection all 50,000 tonnes of steel in the bridge would eventually crumble into a fine red powder.

Why is blood red?

Blood is red because it contains iron.

The iron is found in a protein called haemoglobin, which is found inside red blood cells. These doughnut-shaped cells float in a fluid called plasma, which is carried around your body in the veins and arteries.

Plasma itself is pale yellow, but suspended in it are billions of red blood cells, about 5 million per millilitre of blood. There are about 6.82 litres (12 pints) of blood in your body, making up around 7 per cent of your body weight. Your heart pumps this blood around your body at a tremendous rate; an average red blood cell circulates round the body three times a minute, travelling 19,000 km (12,000 miles) a day – the equivalent of travelling from California to New York and back, twice!

Blue on the outside

It is a common misconception that since veins look blue beneath the skin, they must contain blue blood. It is true that veins carry deoxygenated blood (blood which has given up most of its oxygen and is travelling back to the lungs to pick up some more), and that this is a slightly different colour from the oxygenated blood found in the arteries.

However, the veins only look blue because the colour of light that can penetrate the skin, bounce off the outside of the vein, and travel back through the skin to your eyes is blue light. Deoxygenated blood is actually a dark purplish red, while oxygenated blood is bright red; dark blood exposed to the air will quickly turn bright red as it absorbs oxygen.

Not all blood is red; horseshoe crabs and some other animals have blue blood, while insects may have green blood (although it isn't really blood). Horseshoe crab blood is blue because instead of haemoglobin it has a related protein called haemocyanin, which uses copper instead of iron. In its deoxygenated form haemocyanin is colourless, making the crab's blood a pale greyish colour, but when copper is bound to oxygen it turns blue, giving the horseshoe crab blood a startling blue colour.

Insect 'blood' is not used for oxygen transport at all, so it shouldn't really be called blood – the correct term is haemolymph. If insects don't use blood to carry oxygen, how do their cells get the oxygen they need? Insects have evolved a more direct route for oxygen transport – a system of air passages which reaches into every nook and cranny, so that no cell is far from a direct source of oxygen (this tends to limit the maximum possible size that insects can reach, which is why mammals can be as large as blue whales but insects no bigger than a hand-sized cockroach). Insects use their haemolymph to carry food, and it can be either colourless or it can pick up colour from what the insects eat, which is why vegetarian insects sometimes have green blood.

Why do trees drop their leaves?

Trees drop their leaves because it gets too dry, cold or dark to photosynthesize, so it's not worth keeping them.

Not all trees drop their leaves. Trees that don't drop their leaves are called evergreens. Trees that drop all their leaves at once are called deciduous (from the Latin for 'falling down'). Some evergreens, such as trees in tropical rainforests, do drop their leaves, but not all at once. So really there are two questions here. Firstly, why does any tree ever drop a leaf? Secondly, why do some trees drop all their leaves at once?

Keeping a leaf alive costs energy. But so does letting one drop off, since whatever nutrients are left in the leaf when it falls will be lost to the tree. A tree drops a leaf when it will cost more to keep it alive than to let it fall. This is a bit like having a car; usually it will be cheaper to keep repairing the car you have, rather than buying a new one. But when the car gets really crappy, it may be cheaper in the long run to scrap it and get a new one.

*A tree drops a leaf when it will cost
more to keep it alive than to let it fall*

During dry seasons or cold, dark winters, the cost of keeping a leaf alive goes up, while the amount of photosynthesis it can do goes down. Photosynthesis depends on water and light, and as the temperature drops so does the rate of photosynthesis. For many trees living in the north or south, by the time autumn comes it is simply not worth holding on to their leaves. This is especially true for broadleaf trees, which have big, fragile leaves that can be damaged by frost. So deciduous trees decide to cut and run, or in other words, drop all their leaves at once. A full-grown tree can have more than 200,000 leaves, and over the course of about sixty years will drop over 1.5 tonnes of leaves. Worldwide there are more than 400 billion trees (about sixty for every human being on the planet), so that adds up to a lot of leaves.

When a tree decides that it will be cheaper in the long run to scrap one of its leaves, it drains as many nutrients as possible out of the leaf before cutting it off at the base. The order in which these nutrients are drained is what makes leaves change colour. Leaves are green because they have a lot of chlorophyll to help them harvest blue and red light (see page 13), but they also have other types of pigment to help them harvest other colours of light. These pigments include xanthophylls (which are yellow) and carotenoids (which are yellow, orange and red). Because chlorophyll is the most valuable of these pigments, it gets drained out of the leaf first, leaving the other colours behind, and so leaves turn red, orange and yellow. Eventually even these colours are drained, leaving just tannins, which are brown.

Why do plants flower in the spring?

Many plants flower in the spring so that they can take advantage of good weather to reproduce and still have time to spend the summer making seeds.

Not all plants flower in the spring. At any time of year there will be some flowers that bloom. But spring is the most popular time because plants that flower then get the best balance for their different needs.

The plant's ultimate aim is to reproduce, but for this to happen it has to go through a series of steps: fertilization (which is where the flower comes in), seed production and seed dispersal (which means spreading the seeds to where they can grow). These jobs are much easier when conditions are good – in other words, when there is lots of sunlight and water, the temperature is high and the weather not too wild. In the temperate regions of the world (such as Europe, most of North America, Japan, Argentina, most of Russia and China), the best weather is in summer, and plants time their activities to take advantage of this.

If a plant flowers in winter, it will struggle to get enough energy to build a flower (which is an energy-intensive business); there won't be many insects around to help pollinate the flower (see below); and bad weather, such as high winds, rain and frost, will damage or destroy the delicate flower. If a plant waits until summer, it will have much more energy available for flowering, but it may not leave itself enough time to grow seeds and spread them around before winter arrives again. It seems like the best time for most plants to flower is spring: the weather is getting better, there is sunlight to provide energy and there are plenty of insects around. The plant then has the whole summer to take advantage of good conditions to grow a healthy, successful seed.

Matchmaking

Plants use flowers to attract insects because they need help with pollination. Pollination is the plant equivalent of having sex – the process of getting together the male and female gametes to achieve fertilization. Since plants can't wander around looking for a mate, they rely on insects to do it for them. The insect is lured in by the flower, lands on it and picks up pollen, which it carries off to the next flower.

Flowers may offer special treats to attract the insects, such as nectar. Honeybees collect nectar from flowers and take it back to their hives to make into honey, which they eat during the winter when there are no flowers around. It takes a lot of nectar to make honey: to make 450 g (1 lb) of honey, bees have to fly about 88,513 km (55,000 miles) and tap two million flowers. But it's worth it – just 28 g (1 oz) of honey contains enough energy for a bee to fly around the world.

How do plants know when it's the right time to flower? Some plants seem to measure the temperature before flowering; others rely on the number of hours of daylight. Many species of plants need a cold shock before they will flower, such as the winter frost. There is some evidence that climate change is actually changing the timing of flowering for many plants. According to a study in the journal *Science*, the average first flowering date of 385 British plant species has advanced by four and a half days over the last ten years, and one in seven species is flowering more than a fortnight earlier.

Why do we get old?

We get old because our body's repair mechanisms stop working as well and we start to break down like old cars.

Actually this is more a description of how we get old. Nobody really knows why we get old. Not all animals age so it may be possible for humans not to.

Various theories in biology concern ageing. Some people think that our bodies can soak up only so much damage before the repair mechanisms that mend this damage are exhausted and stop working. Another theory concerns the ways our cells avoid turning into cancer cells: it may be that the same mechanism that helps a young person avoid getting cancer – making cells kill themselves before they become dangerous – eventually causes the same person to get old. This is called the antagonistic pleiotropy theory. A third theory, called the disposable soma theory, says that our bodies exist only to get our reproductive cells into the next generation, like a messenger whose only job is to deliver a message. It doesn't matter if the messenger (or, in biological jargon, the soma) gets damaged or killed, so long as he or she delivers the message first. In other words the messenger or soma is disposable.

These types of theory all say that ageing has no purpose, that we get old because it is impossible not to get old. But some animals don't seem to get old. The Aldebra giant tortoise can live to at least 255 years, and may be able to go on forever. The same is true of the rougheye rockfish (oldest-known individual 205 years) and even the humble lobster. The quahog clam can live for over 400 years! The oldest animal ever discovered is a quahog clam nicknamed Ming, which was over 405 years old when it was fished out of the ocean near Iceland.

Aldebra tortoise

To take into account the ability of some animals to live without ageing, some biologists have suggested that ageing does have a function, and that we have actually evolved to grow old. Age and death might help clear the way for our descendants to be more successful, or it might be that ageing poses such a survival challenge that it acts as a kind of test, weeding out weaker individuals so that only the strongest survive. If we really have evolved to grow old, then the ageing process must be under the control of our genes, which in turn means we can do something about it. Advances in genetic science and the ability to modify our genes (genetic engineering) might make it possible for future humans to stay young forever.

Why do we feel pain?

We feel pain so that we know to stop doing whatever is causing the pain.

Pain is your body's way of signalling to your brain that it is being damaged. If something causes damage to the cells of, say, your foot, pain receptors in your foot fire off nerve signals that travel along your nerves to your brain. Your brain translates these signals into the feeling of pain, and this makes you do something to stop the pain. If the pain had started after you stepped on a very pointy rock, you might take your foot off it. Without pain, you wouldn't know that you needed to move your foot, and so you might end up losing it, which wouldn't do your chances of survival any good at all.

People who cannot feel pain don't live very long. There are some conditions, such as syringomyelia, which stop the pain receptors from working properly, and as a result people don't feel pain. This can be dangerous in lots of ways – if you don't realize you've cut your foot, you won't do anything to stop it getting infected (like keeping it clean). But the greatest danger to people who can't feel pain is joint damage. If you hold one position for too long, your joints will start to hurt, prompting you to change position before you've done any damage to your joints. People with syringomyelia don't change their position because they don't realize they are damaging their joints, and they end up destroying them. Eventually they die of blood poisoning.

Feeling pain: from trigger to perception

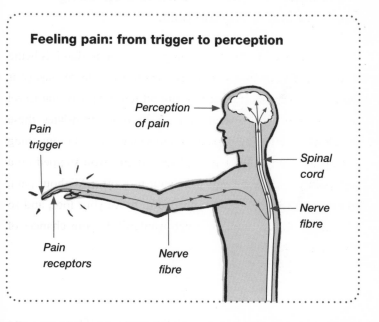

Pain can be hard to describe because everyone's experience of it is different. Dr Justin Schmidt studies biting and stinging insects and he has devised a scale that rates and describes the pain caused by different bugs. At the bottom of the scale, with a pain rating of 1.0, is the sweat bee; Schmidt describes its sting as 'light... almost fruity... [as if] a tiny spark has singed a single hair on your arm.' Near the middle of the scale comes the bullhorn acacia ant with a score of 1.8; its bite feels like 'someone has fired a staple into your cheek'. The most painful insect is the pepsis wasp: its sting scores 4.0 and Schmidt says it is like 'a running hair dryer has been dropped into your bubble bath'.

Why do we sleep?

We sleep to recharge our batteries and so that our brain can rewire itself.

It used to be thought that there was no special reason for sleep, and that it was just what happened when animals stopped moving to save energy and keep out of trouble. But we now know that all animals sleep, even ones that live in the sea and must keep moving to survive. We also know that animals that don't get enough sleep suffer and eventually die, so there must be a good reason for it. In other words, sleep does have a purpose and function, but we don't know what it is yet.

Wear and tear

There are many different theories for why we need to sleep. Animals that are not allowed to sleep get sick and die: their wounds don't heal, they lose weight, their skin gets bad and their immune systems break down. So one reason we need to sleep is probably so that the body can repair and look after itself in a way that it can't while awake.

The longest a human being has gone without sleep is two hundred and sixty-four hours (eleven days), although there are claims that a British woman stayed up for eighteen days and twenty-one hours during a rocking chair marathon. People who go for so long without sleep have hallucinations (sense things that aren't there) and memory loss, become paranoid, confused and bad-tempered, and cannot do simple tasks such as adding up or counting. Not getting enough sleep over a long period interferes with memory and stops people from learning. Researchers have discovered that sleeping brains have bursts of activity. This has led scientists to suggest that one function of sleep is to give the brain a chance to reorganize itself, file away all the information picked up during the day and do other tasks that it can't do while awake.

Possibly the sleepiest animal in the world is the koala bear, which sleeps about twenty-two hours a day. Male lions supposedly sleep for up to twenty hours a day (leaving female lions to do most of the work). Arguably the least sleepy animal is the dolphin, because it always keeps one side of its brain awake (and one eye open) – probably because it might drown otherwise.

Why do we have forty-six chromosomes?

Humans have forty-six chromosomes because the great apes, with whom we share common ancestors, have twenty-four pairs of chromosomes, and some time after humans split from the great apes two of these chromosomes got joined up.

Humans have two copies of each of twenty-three chromosomes, giving forty-six in total. Great apes, such as chimpanzees and gorillas, have two sets of twenty-four chromosomes, giving forty-eight in total. Around 6–8 million years ago humans and great apes shared an ancestor, an ape with two sets of twenty-four chromosomes. But after our species split from the apes, two of these twenty-four chromosomes got stuck together to form the human chromosome 2. Chromosome 2 looks like two chimpanzee chromosomes joined together, because in evolutionary terms that is what it is.

Chromosome fusing is actually surprisingly common. About one in a thousand babies have two chromosomes stuck together, and it usually makes no difference to their health (although half of their eggs or sperm will be infertile). The same process has happened to other animals. Wild horses have thirty-three pairs of chromosomes (giving sixty-six in total), but domesticated ones have thirty-two pairs (giving sixty-four in total) – in other words, horses have recently undergone the same chromosomal fusing that humans had millions of years ago.

The twenty-three pairs of chromosomes of a human male

1   2   3   4   5   X

6   7   8   9   10   11   12

13   14   15   16   17   18

19   20   21   22   Y

Humans have one fewer pairs of chromosomes than great apes

Sex chromosomes determine gender (X for female and Y for male)

Mr Potato

Humans have forty-six chromosomes because our ape ancestors had forty-eight, but why did they have forty-eight? Nobody knows. There seems to be no special importance to the number of chromosomes a plant or animal has. We have the same number of chromosomes as the potato. Chickens have seventy-eight, a goldfish has ninety-four and there is a type of fern (Ophioglossum reticulatum) *that has one thousand two hundred and sixty.*

The number of chromosomes an animal or plant has is probably down to random mistakes. When an animal or plant makes special cells for reproduction (called gametes), it usually splits its sets of chromosomes in half. In humans, each sperm or egg cell gets just one copy of each of the twenty-three chromosomes. When a sperm and egg come together at fertilization, the fertilized egg then has forty-six chromosomes. But the chromosome sorting process often goes wrong, and it is easy to end up with a sperm or egg that has two copies of one of the chromosomes, or even all forty-six chromosomes. In humans such gametes cannot usually make a baby, but in some other animals and especially in plants this doesn't seem to be a big problem. When you add in the possibility of chromosomes fusing together, as has happened with humans and horses, you can see that it is easy to end up with any number of chromosomes. It seems to be down to dumb luck.

Why do we have X and Y chromosomes?

We have X and Y chromosomes because about 170–310 million years ago what used to be a second X chromosome lost most of its genes, along with the ability to swap genes with the other X chromosome, and became the Y chromosome.

Many other animals, including the ones we evolved from around 170–310 million years ago, have two X chromosomes and no Y chromosome, so perhaps a better question would be 'Why do some humans have a Y chromosome?' The answer is, like most species of animal, we need a way to make babies of different genders (i.e. both male and female), and the Y chromosome is what humans have evolved to achieve this. In humans, whether you are male or female is decided by whether or not you have a Y chromosome.

We all have twenty-three pairs of chromosomes, giving us forty-six chromosomes in total (see page 63). Out of these twenty-three pairs, twenty-two are identical copies of each other, so that a person has two copies of chromosome 1, two copies of chromosome 2, etc. The only exception to this is the pair known as the sex chromosomes (they are called this because they decide the sex of the person). These come in two different flavours, called X and Y, because they look a bit like the letters X and Y when seen under a microscope.

The Y chromosome carries the genes that make a baby turn out to be a boy. So if a baby grows from an egg that has one copy of the X chromosome and one Y chromosome, it will turn out to be a boy. If the baby has two X chromosomes and no Y chromosome, it will not have the 'male' genes and will develop into a girl.

The Y chromosome used to be an X, and the very early mammals all had two X chromosomes. In other words they didn't have sex chromosomes. They must have had a different way of deciding which babies would be male and which female, just as fish, birds and reptiles do today. Some animals, such as alligators and turtles, use temperature to make an egg turn out male; others use hormones. Some fish, such as clownfish (the fish from *Finding Nemo*), can even change sex after they are fully grown. If there are too many males and not enough females, a male clownfish can change into a female.

Why did the dinosaurs die out?

The dinosaurs probably died out because of a combination of natural disasters: global warming over millions of years, followed by a series of colossal volcanic eruptions in India that released vast quantities of poisonous gas, at almost the same time as a massive asteroid smashed into the Earth.

The dinosaurs ruled the planet for over 160 million years, but 65 million years ago something terrible happened that wiped out every species of dinosaur (except for the ancestors of birds).

So what was this terrible event? Most people nowadays have heard that an asteroid or meteorite killed the dinosaurs (a meteorite is an asteroid that has fallen to Earth), and there is a lot of evidence that a 10 km (6.21 miles) wide asteroid slammed into Mexico about 65 million years ago. The crater it left is hidden under rocks and water, beneath an area called Chicxulub, and the event is known as the Chicxulub impact.

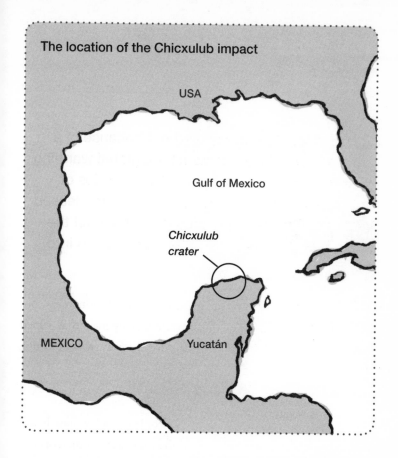

The location of the Chicxulub impact

USA

Gulf of Mexico

Chicxulub crater

MEXICO

Yucatán

The Chicxulub impact blasted a crater 180 km (111.85 miles) wide, setting off earthquakes around the planet. The explosion took out most of North America almost instantly with a flash of intense heat and a destructive shockwave, battering the rest of the planet with super-powered hurricanes – or hypercanes – five times more powerful than the biggest hurricanes we get today. A huge wall of water known as a tsunami raced outwards from the impact site, reaching up to 300 m (about 328 yd) high and sweeping 300 km (about 186 miles) inland. But all this was only the beginning.

Vast amounts of melted rock were thrown into the sky and rained down all over the planet, setting fire to everything. Up to a quarter of all life on land was burned to ash. The dust and smoke shut out all sunlight, probably for several years, so that the Earth froze in the darkness as temperatures fell by up to 15°C (59°F). Plants could not photosynthesize, and poisonous gas made the oceans toxic. The ozone layer was damaged, so that when the smoke finally cleared the Earth's surface was bathed in deadly ultraviolet light.

As if this were not bad enough, life on Earth was probably already suffering as the result of a series of volcanic eruptions in India known as the Deccan Traps. These were not like ordinary volcanic eruptions; they were much, much worse. About 512,000 km³ (about 122,835 miles³) of molten rock poured out of a titanic gash in the Earth's surface. When it cooled it left a layer of solid rock over 2 km (1.24 miles) thick in places, covering an area of 500,000 km² (about 193,051 miles²).

The final blow

The heat, ash and poisonous gas released by the Deccan Traps eruptions helped push the dinosaurs close to extinction. The Chicxulub impact was probably what finished them off.

Humanity could easily go the way of the dinosaurs. There are several places on Earth, such as Yellowstone National Park in the US, and the area around Naples in Italy, where eruptions on the scale of the Deccan Traps could happen at any moment. Also, there are millions of asteroids that cross the Earth's orbit, and some of them could be 10 km (6.21 miles) wide or bigger. Over a long time scale, your chances of dying because of an asteroid impact are roughly equivalent to your chances of dying in an air crash.

Why are we running out of oil?

We are running out of oil because we use it up millions of times faster than it can form.

Oil is formed when tiny plants and animals in the sea die, sink to the bottom of the ocean, get buried under layers of rock and are crushed and heated until they break down into very simple molecules. This process takes thousands of years, at least – it may take millions of years. So for all intents and purposes, oil is a non-renewable resource. This means there is only so much in the ground, and once we suck it out and use it, it is gone.

If you keep using a non-renewable resource, eventually you will run out because nature is not making any more of it. How soon you will run out depends on how quickly you are using the resource, and as far as oil goes, we are using it very quickly indeed. Global oil consumption hit a new peak in 2010, at 87.4 million barrels per day. This means the world burns through over a thousand barrels of oil per second.

According to the annual 'Statistical Review of World Energy' produced by oil giant BP, the world's oil reserves at the end of 2010 were estimated to be 1.383 trillion barrels. This means that if oil use continues at the same rate as in 2010, the world would run out of oil in just over forty-three years. In fact, the rate of oil consumption continues to rise every year, so the global oil reserve as estimated by BP should be depleted even sooner than this.

But the picture is not that simple. On the one hand there are many who argue that the estimates of how much oil is left are far too optimistic, with oil-producing countries such as Saudi Arabia claiming to have far larger reserves than they actually do. On the other hand there are many others who argue that we are underestimating oil reserves, as it will be possible to find or produce new types of oil.

An example is the development of oil sands as a source of oil. Oil sand is black and sticky with tar; by washing off the tar, collecting it and refining it, it is possible to get oil. Because this is difficult and expensive, oil sands were not previously considered to be a realistic source of oil, but now things have changed (specifically, the price of oil has gone up, making it worthwhile). Canada is now considered to have the third-largest oil reserves in the world, thanks to the 169 billion barrels of oil that can be extracted from its oil sands. As technology improves, there is a good chance that 'official' estimates of oil reserves will get bigger, so that we may not run out of oil for fifty or a hundred years.

The picture is even more complicated than this. Some experts say that the question is not 'when will the world run out of oil?', but 'when will we need more oil than we can produce?' This moment, when demand outstrips supply, is called the 'peak oil' moment, and according to peak oil theory it will lead quickly to global disaster as oil prices rocket, the world economy collapses and civilization itself is threatened. For instance, over 90 per cent of transport in the world depends on oil – if it suddenly becomes incredibly expensive, almost no one will be able to afford to fly or drive anywhere. Peak oil theory is not widely accepted, for many of the same reasons that oil reserves are thought to be sufficient for the foreseeable future.

Why does the sea appear blue?

The sea appears blue because water absorbs more red and yellow light than blue light.

A glass of water looks colourless, but in fact it is very, very light blue. If you have enough water in one place, as in the deep sea, this blue colour becomes visible. But the sea is not always blue. Close to the coast it often looks green, while it is also possible to get red, yellow and brown seas. These colours depend on what is in the water, and which colours of light the water absorbs and which it reflects.

Sunlight falling on the sea is made up of all the colours of the spectrum (i.e. it is basically white light). Some of the sunlight reflects back off the top of the water; the rest of it passes into the water. The water molecules absorb certain colours or wavelengths of light more than others. The colours they absorb most are red, yellow and green, and this leaves blue light. Some of this blue light keeps going down into the water, and some is scattered by the water molecules so that it comes up out of the water and meets your eyes.

About 65 per cent of the visible light entering the ocean is absorbed within the first metre of water; less than 1 per cent penetrates as far as 100 m (109.36 yd), and this is entirely blue light.

Muddying the waters

Close to the coast, there tends to be a lot more sediment (small particles of dirt) in the water because of all the dirt that washes off the land into the sea. This sediment changes the way light is absorbed and reflected; the floating specks of dirt tend to absorb red and blue light the most, leaving green light to reflect back out of the water and making coastal waters appear green.

Various other colours can be caused by tiny plants floating in the water. Known as phytoplankton, these tiny plants use pigments such as chlorophyll to harvest sunlight, and they can stain the water. Green phytoplankton make the water look green, but phytoplankton comes in many different colours, depending on the pigment used by each species. Some phytoplankton species use red pigments, and these are responsible for what are called red tides.

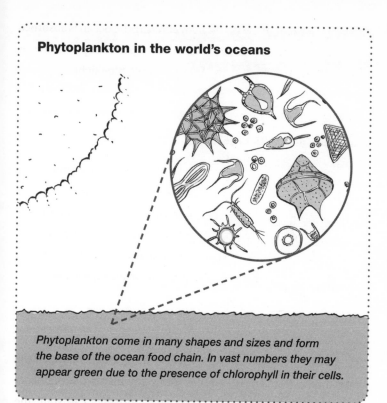

Phytoplankton in the world's oceans

Phytoplankton come in many shapes and sizes and form the base of the ocean food chain. In vast numbers they may appear green due to the presence of chlorophyll in their cells.

The seabed also affects water colour if the sea is shallow enough for light to reach and reflect off it. In tropical areas with clear, shallow water, the ocean appears very light blue because of the colour of the white sandy bottom mixing with the blue water.

Water covers 71 per cent of the Earth's surface and this is why the Earth appears blue from space.

Why does radioactive material have a half-life?

Radioactive material has a half-life because this is how we describe the rate at which the material undergoes radioactive decay.

Radioactive decay is the process by which a radioactive element, known as a radionuclide, changes into a different radionuclide by shooting out subatomic particles or energy. The giving off of subatomic particles or energy is known as radiation.

The half-life of a radionuclide is the length of time it takes for half of the atoms in a sample to decay. For instance, bismuth-212 has a half-life of just over an hour. This means that if you started with one thousand atoms of bismuth-212, in an hour there would be about five hundred atoms left. The others would have decayed to a different radionuclide. After another hour there would only be around two hundred and fifty atoms of bismuth-212, and after three hours there would be one hundred and twenty-five.

An analogy is with flipping coins. Imagine that a flipped coin that comes up tails is like a radioactive atom that has decayed, and then imagine you have a thousand coins and once an hour you flip them all and then throw away the ones that turned up tails. After one hour you would have roughly five hundred coins, after two hours you would have two hundred and fifty, and after three hours you would have one hundred and twenty-five.

Why not just have a number that tells you how long it will take for a radioactive atom to decay? Such a number might be called the 'lifespan' of a radionuclide. The reason that we use half-life instead of lifespan is that radioactive decay is a completely random process. It is controlled by something called the uncertainty principle, which means that it is impossible to say for certain exactly when a radioactive atom will decay. It is only possible to state the probability of that decay.

For an individual atom of bismuth-212, it is possible to say that over the course of an hour there is a 50 per cent probability that it will decay, but there is no length of time over which it is possible to say with 100 per cent certainty when it will decay. Although it is astronomically unlikely, it is possible that any single atom of bismuth-212 still won't have decayed in a thousand years. So we have to use a measure like half-life, which involves probability.

Bismuth-212 has a relatively short half-life, but the radionuclide with the shortest half-life is helium-5, with a half-life of just 7.6 x 10^{-22} seconds, which is less than a billion trillionth of a second. The radionuclide with the longest half-life is tellurium-128: 2.2 x 1024 years (2.2 trillion trillion years).

Why does the wind blow?

The wind blows because some parts of the Earth's surface heat up more than others, causing air to rise, which draws in air from elsewhere.

Wind is the movement of air over the Earth's surface. Air moves from one place to another to even out differences in pressure. If you are in London or New York, there is about 12 km (7.46 miles) of air piled up on top of you, and this mass of air presses down with an average pressure of 1 bar (1,000 mb) or 1 atmosphere. Where part of the Earth's surface is hotter, the air above the surface is heated, and hot air rises. Rising air creates low pressure. Gases, such as air, flow from areas of high pressure to areas of low pressure.

Why do some parts of the Earth's surface heat up more than others? Latitude (position relative to the equator and the poles) is a big influence because the equator gets much more concentrated sunlight than the polar regions. Land and sea are other major influences: land heats up and cools down much faster than water, so that during the day the land tends to be hotter than the sea and vice versa at night. Other influences include ocean currents, forests and mountains.

Rise and fall

Heating at the equator creates a permanent band of low pressure around the Earth, known as the doldrums. Here air is rising to the top of the atmosphere and spreading out. Meanwhile at what are known as the horse latitudes, air is falling back to Earth creating bands of high pressure. This air flows back toward the equator as the trade winds.

Global wind patterns

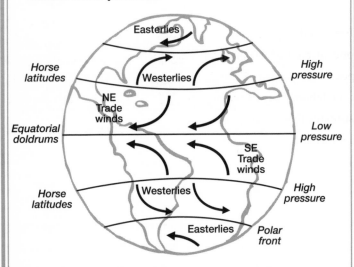

At the horse latitudes westerlies blow towards the poles and the trade winds (winds blowing from the east) flow back towards the equator.

The rotation of the Earth changes the direction of the wind and causes systems of high and low pressure to spin in enormous whirls, known as cyclones and anticyclones. In tropical regions these systems can turn into hurricanes, where the warm ocean pumps so much energy into the atmosphere that wind speeds reach up to 120 kph (74.56 mph) or more. Hurricanes contain terrifying amounts of energy: in a single day a hurricane releases enough energy to power the entire US for six months, or provide a whole year of power for a country like Britain or France.

Over land, special types of storm cloud can create very concentrated whirls of wind called tornadoes. The highest wind speed ever measured was 516 kph (318 mph) during a tornado in Oklahoma in 1999, but the US National Weather Service estimates that wind speeds can reach up to 800 kph (500 mph) in a tornado.

Why is the Earth round?

The Earth is round because gravity pulls with equal strength in all directions, so that any uneven bits are pulled back in line with all the other bits, making a sphere.

Imagine a crowd of people standing in a field, each person tied to the others. Now imagine they all start pulling on the ropes that bind them together. The crowd of people will quickly be pulled into a circle, and anyone who tries to move out of the circle will be pulled back in. This, in three dimensions, is what happens to the Earth – it's called hydrostatic equilibrium.

This is what tries to happen with any and all mass, it's just that gravity is a very weak force. Your mobile phone, for instance, exerts gravity on itself and everything around it, but its gravitational pull is so incredibly tiny that you can't feel it and it cannot possibly overcome the strength of the metal and plastic that gives the phone its rectangular shape. But the larger an object gets, the stronger its gravity. Any asteroid bigger than about 1,000 km (621.37 miles) across will collapse into a sphere under its own gravity. This is now one of the definitions of a planet, according to the International Astronomical Union.

In fact, the Earth is not completely round. Once every twenty-four hours it spins round an axis that runs through the poles, and this means that the equator is moving much faster than the poles. With the surface of the Earth at the equator moving at 1,675 kph (1,040 mph), there is a centrifugal force pressing out, and this makes the Earth bulge at the equator. The distance from the centre of the Earth to the equator is about 0.33 per cent further than the distance from the centre of the Earth to the poles, which means that the diameter of the Earth from pole to pole is 43 km (26.72 miles) less than across the equator. So the top of Mount Everest is not the point on the Earth's surface furthest from the centre of the planet – that's actually Mount Chimborazo in Ecuador, which is almost on the equator. Being thicker in the middle means the Earth is not a sphere, but an oblate spheroid.

As well as this bulge around the middle, the Earth also has mountains, valleys, ocean trenches and other wrinkles, so its surface is not smooth. But compared to the overall size of the Earth these wrinkles are tiny. A billiard ball is smooth to within a tolerance of 0.22 per cent, but the Earth is smooth to within a tolerance of 0.17 per cent, so if a galactic-size giant came along and handled the Earth it would feel smoother than a billiard ball.

Why is there a dark side of the moon?

There isn't.

All parts of the moon, with the exception of the corners of some deep craters that are in permanent shadow, are illuminated by the sun half the time.

There is, however, a far side of the moon, which Earth-based observers never get to see, and which remained one of the great mysteries of nature until a Russian spacecraft orbited the moon in 1959.

Why do we only get to see one side of the moon? The moon is in a synchronous orbit around the Earth; in other words, as the moon goes around the Earth it also rotates on its axis at exactly the right speed to ensure that the same side is always facing the Earth.

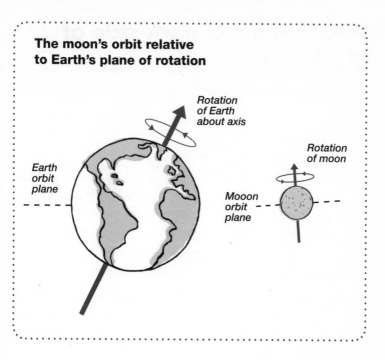

The moon's orbit relative to Earth's plane of rotation

Rotation of Earth about axis

Earth orbit plane

Rotation of moon

Mooon orbit plane

This was not always the case. The moon was probably formed about 4.5 billion years ago, when a planet the size of Mars crashed into the young Earth and was smashed apart. The rubble went into orbit around the Earth and collected together into the moon, which with a diameter of 3,476 km (2159.89 miles) is just over a quarter the size of the Earth.

At this time the moon was much closer to the Earth and both spun much faster, but the effects of the tides (see page 22) caused them to slow down and move further apart (today they are 384,400 km (about 238,855 miles) apart, a distance that increases by 3.8 cm (1.5 in) a year. If there were no moon the Earth would now be spinning about three times faster, giving us 1,095 days in a year, each of them just eight hours long.

The same 'braking' effect acted on the moon, slowing its rotation until it reached what is called the 'tidal locking point', when its rotation became synchronous (perfectly coinciding) with its orbit, so that the same side always faces Earth.

Seeing in the dark

It could be said that the far side of the moon is 'dark' in the sense of invisible and unknown, except that this is no longer true either. Spacecraft have now photographed the far side of moon extensively, starting with the Soviet probe Luna 3 in 1959. In addition the moon wobbles slightly as it spins, making a few degrees of the far side visible from Earth.

So what's on the far side of the moon? Mainly a lot of craters named after Soviet heroes because it was their scientists who got to assign names first, giving us features such as the Mendeleyev and the Korolev craters.

The far side is quite different from the near side, with fewer 'seas' (large, dark, relatively smooth areas formed from floods of lava) and more rugged, cratered highlands. This is because the far side has a thicker crust than the near side, a discrepancy which is another of the great mysteries of nature.

One recent theory is that Earth originally had two moons, which collided around 4.4 billion years ago. The smaller one was smeared or 'splatted' across the face of the moon that would become the far side, and the impact drove the young moon's magma across to the near side, where it produced a thinner crust and more frequent lava-flooding events.

Why can't we breathe underwater?

We can't breathe underwater because our lungs cannot extract enough of the oxygen that is dissolved in water.

Fish can breathe underwater so why can't we? Also, why can't fish breathe when they are out of the water? Fish use gills, which are similar to lungs in both form and function. If gills and lungs are similar, why are we and fish stuck in our different worlds?

The answer is that gills and lungs are each specialized for their particular environments. Seawater contains 1.5–2.5 per cent dissolved air, and about a third of this is oxygen. This turns out to be plenty for fish, which are cold-blooded and have low oxygen needs compared to warm-blooded animals like mammals and birds, with their faster metabolisms. Fish evolved gills to help them extract dissolved oxygen from water: gills have lots of tiny blood vessels very close to the surface of the skin, and skin that is so thin it is called a membrane. Oxygen from the water can easily pass across the gill membrane into the blood, and waste gases such as carbon dioxide can pass out.

Gills work fine when they are wet, and in fact a fish could breathe on land if there were some way to keep its gills constantly refreshed with water. When animals left the sea and moved onto land they couldn't manage this trick, and so they needed a different system: the lungs. With lungs, the whole process relocates to inside the body, where the lung membranes can be kept wet, protected and warm. With lungs, land animals have been able to take advantage of the much higher level of oxygen in the atmosphere than in water – at sea level the air is about 21 per cent oxygen, offering up to forty times more oxygen than water. Since oxygen is needed to power the muscles, the brain and the rest of the body, air-breathing animals can be much more energetic than their water-breathing counterparts. This is why mammals that have gone back to the sea, like whales and dolphins, have kept their lungs and still need to come to the surface to breathe air.

Below the surface

Lungs won't work underwater because getting filled up with water damages the delicate set-up inside the lungs, but even if we had gills we wouldn't be able to breathe underwater because there simply isn't enough oxygen in the water to meet our needs.

The closest a human can get to the experience of breathing underwater is to breathe perfluorocarbon, a liquid which can dissolve high levels of oxygen. Breathing liquid would make it possible for divers to go much deeper because they wouldn't have the problems with pressure that breathing compressed air brings. James Cameron's film *The Abyss* memorably features a character who breathes perfluorocarbon in order to descend to great depths underwater.

Why does time go forward?

Time goes forward because you can't unspill milk or unshatter a glass.

One of the great mysteries of physics is why time's arrow points forward. The arrow of time is an idea thought up in 1927 by British physicist Arthur Eddington, who used it to describe how time seems to move in only one direction. You can move from the past into the future, but not back the other way. If space and time together make four dimensions, time is the only dimension in which you must travel in one direction.

All of the equations that physicists use to describe the universe, such as ones dealing with gravity and electromagnetism, work perfectly well in either direction. They are said to be symmetrical, and they don't seem to be affected by the direction of time's arrow. Yet time does have an arrow – why should this be?

The answer is the second law of thermodynamics, which says that entropy always increases. Entropy means disorder, or unusable energy. In a closed system, where you are not allowed to add any energy to what you started with, disorder will always increase. In real life this means that if you spill milk from a jug the milk will not leap back into the jug, or if you drop a glass and it shatters, it will not reassemble itself and leap back into your hand. If you ran a film of either of these events you would be able to tell whether it was running backwards or forwards, unlike the symmetrical equations mentioned above. This increase of entropy means that time can run in only one direction.

The irreversible nature of the second law of thermodynamics is probably linked to the origins of the universe itself, which started off incredibly small and dense and has been expanding ever since. Immediately after the big bang, the universe had incredibly low entropy, and as time goes on its entropy increases. This is called the cosmological arrow of time.

A possible fate for the universe is that it will eventually become completely disordered with maximum entropy, which means that all matter and energy will become evenly distributed. The available energy would end up spread very thinly across the entire universe, so it would be very cold everywhere, and this fate is sometimes called the heat death of the universe. An alternative fate is that the universe will stop expanding and go into reverse, a scenario called the big crunch. In the big crunch entropy might decrease and time might change direction!

Why does the earth quake?

The earth quakes because the surface of the Earth is made up of plates that move and grind against one another, and sometimes they stick and then suddenly slip.

Earthquakes are caused by sudden movements of the tectonic plates, giant slabs of rock that make up the outer skin of the planet. In some places these plates slide past each other but since they are made of rock they don't slide smoothly. Instead they get stuck fast at some points, until strain and stress builds up so much that all at once they give, jumping past each other. When this happens a lot of energy is released and movement ripples out through the earth. On the surface these ripples cause earthquakes.

Earthquakes can also be caused by hot molten rock rising to the surface in a volcano, and by other catastrophic events such as asteroid impacts (see page 68).

The tectonic plates float on a layer of partially liquid rock called the mantle. This moves very slowly – about one ten-thousandth the speed of the hour hand on a clock. As a result the tectonic plates themselves generally move equally slowly – North America and Europe, for instance, each on different tectonic plates, are moving no faster than a fingernail grows.

But the forces involved in moving billions of tonnes of rock are colossal, and these are magnified when the plates stick and tension builds up. The size of the quake that results when the tension is released depends on how long it has been building. This size, or magnitude, is measured with the Richter scale.

Great shakes

The Richter scale is a logarithmic scale, which means that each point on it is ten times greater than the one before. So a magnitude 2 earthquake is ten times bigger than a magnitude 1 quake, but 100 times smaller than a magnitude 4 quake.

Low-magnitude quakes are amazingly common. A quake of less than magnitude 3.4 happens over 800,000 times a year, but is too small for anyone to notice. Magnitude 4.3–4.8 quakes will rattle the houses of a window; they happen nearly 5,000 times a year. It takes a quake of 6.2 or greater to be really dangerous, and 7.4 to collapse a lot of buildings; fortunately such a big quake only happens about four times a year.

The amount of energy released by a big quake is terrifying. The energy released in the San Francisco earthquake of 1904 was comparable to one of the largest nuclear bombs ever created exploding under the city, but this was only around magnitude 7.8. The biggest quake ever recorded was the southern Chile quake of 1960, which registered 9.5 on the Richter scale. The Tohoku earthquake that caused the devastating tsunami in Japan in March 2011 was magnitude 9.0.

Why do light bulbs light up?

Light bulbs light up because electrons have to squeeze through their narrow filament, forcing them to bash into the atoms of the filament, which makes them vibrate and give off heat and light.

An electrical circuit with a battery, wires and a light bulb is like an aquarium with a water pump, tubes and a filter. The battery acts like a pump, forcing water around the tubes and through the filter, which is like a light bulb. The tubes let the water flow quite easily, but the filter is narrow and partially blocked, so it is harder for the water to get through. Squeezing through causes the water to lose energy by friction.

In an electrical circuit the electrons that carry the current are like the water. They can pass through the wires quite easily because they are made of copper and have a low resistance to electricity, but the filament is made of tungsten, a metal with a high resistance. Also the filament is extremely long and thin – in a typical 60-watt bulb the tungsten filament is 2 m (2.19 yd) long but only a quarter of a millimetre thick, which makes it even more resistant. Resistance is a measure of how easily electricity can pass through something. The electrons work so hard to get through the tungsten that they pass energy to the atoms of the filament. As the tungsten atoms build up energy they get excited, and when they get excited they give off heat and light energy.

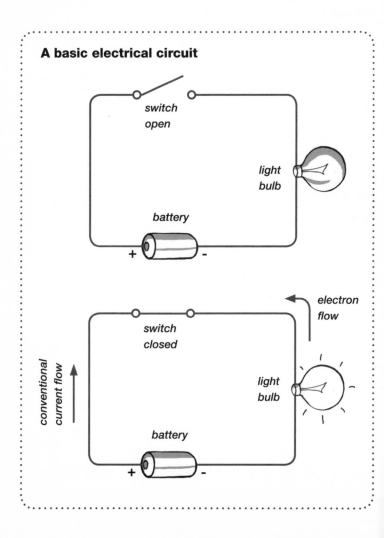

A basic electrical circuit

switch
open

light
bulb

battery

+ −

switch
closed

electron
flow

conventional
current flow

light
bulb

battery

+ −

Like most metals, tungsten has to be white hot before it gives off much light, so light bulb filaments get to 2,200°C (3,992°F). Even then, about 90 per cent of their energy is given off as invisible infrared light (heat), so normal light bulbs aren't very efficient at turning electricity into light.

They also tend to burn out quite quickly so they need replacing often; America alone spends about $1 billion buying 2 billion light bulbs a year – that's 5.5 million a day! Light bulbs can last a very long time, however. The longest-burning light bulb is the Livermore Light in Firestation 6, Livermore, California, which has been burning since 1901. So far it has used up about as much energy as you would need to run a tumble dryer for four years.

Why split the atom?

The atom is worth splitting because it has the highest energy density of any fuel available.

Burning 1 kg (2.2 lb) of coal can power a 100-watt light bulb for about four days, and 1 kg (2.2lb) of uranium can power the light bulb for just over 140 years. This means that, kilogram for kilogram, uranium provides 16,000 times more electricity than coal. Another way of saying this is that the energy density of uranium is 16,000 times higher than coal.

The reason that uranium has such a high energy density is because it is relatively easy to split an atom of uranium with a process known as nuclear fission. When an atom of uranium-235 splits into an atom of barium and an atom of krypton, a minute part of the original uranium atom is converted into energy. Einstein's famous $E=mc^2$ equation tells us that even a tiny bit of mass equals a colossal amount of energy (see page 27).

Splitting the atom releases so much energy that from a single gram of uranium it is possible to make 1 megawatt of energy per day. You would have to burn 3 tonnes of coal or 600 gallons of oil to get the same amount of energy.

The enormous energy density of uranium has important consequences for the environment and energy security. Because fission produces no direct emissions (carbon dioxide, sulphur dioxide, nitrogen oxides, etc.), nuclear energy does not add to greenhouse gases. It does produce some radioactive waste, which poses its own special problems, but the amount of waste produced is relatively tiny. The waste per kilowatt-hour of electricity produced by a nuclear power plant is around 3 g (0.11 oz); for natural gas the figure is 181 kg (399.04 lb), and for coal a staggering 1,064 kg (1.06 tonnes).

Also, because so little uranium fuel is needed to keep a nuclear plant running, it is easy for a country to get enough to supply most of its energy needs. Only 200 tonnes of milled uranium are needed to keep a 1,000-megawatt nuclear reactor going for a year, so a single warehouse can hold enough fuel to guarantee the energy security of an entire country for years. Since one of the main causes of warfare, conflict and terrorism in the world is control of energy supplies, specifically oil, switching to nuclear could make the world a much safer place.

Nuclear future

Some countries rely heavily on nuclear power. Worldwide, nuclear power through fission supplied about 13.5 per cent of the world's electricity production in 2010, but in some countries the figure is much higher. France gets 77 per cent of its energy from nuclear.

The other reason to split the atom is to produce incredibly powerful and destructive weapons. The atomic bomb dropped on Hiroshima in 1945 was 2,000 times more powerful than the biggest bomb ever used up until then. The largest nuclear weapon ever created, the Soviet Tsar Bomba of 1961, was ten times more powerful than all of the explosives used in World War II put together.

Why can't we eat grass?

We can't eat grass because we don't have bacteria in our guts that can break down tough cellulose.

In fact, we do eat grass – lots of it. About three quarters of all the food humankind eats comes from grasses, specifically wheat, rice and corn. But the bits we eat are the seeds, not the green, leafy stems. Grass stems and similar green leafy bits of plants make up the major part of the diet of many common animals, including cows, sheep, kangaroos and horses. If they can eat it, why can't we?

The answer lies in one of the main differences between grass and us. The cells that make up our bodies are relatively flimsy, since they are protected only by a fragile membrane made of a thin layer of fat. The cells that make up plants are much stronger, because as well as a membrane they have an extra layer of protection: a cell wall made of cellulose, a tough, starchy substance. Humans cannot digest cellulose, which is why we can't live off grass. Sellotape, the well-known clear sticking tape, is made of cellulose, which should give you an idea of how inedible it is.

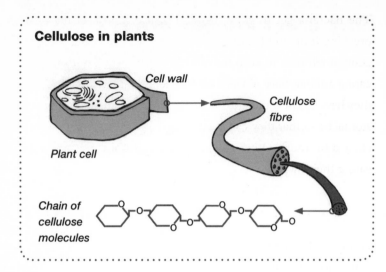

Cellulose in plants

Cell wall

Cellulose fibre

Plant cell

Chain of cellulose molecules

Cellulose is a very long molecule formed by linking many small sugar molecules together. Humans can digest sugar easily enough, so why can't our guts cope with cellulose? In order to break up cellulose and split it into simple sugars, an enzyme called cellulase is needed. Unfortunately we don't make cellulase.

Cows and other grass-eating animals don't make cellulase either, but they have a lot of tiny friends who do – bacteria that live in their guts. These special bacteria live in a symbiotic relationship with the animals, which means that each of them benefits. The bacteria get a nice warm place to live and a constant supply of chewed-up grass to feed on, and in return the animal takes advantage of the enzymes that only the bacteria can produce.

The bacteria get to work on the tough cellulose in the grass, breaking it up and feeding off some of the sugars, but there are enough left over to keep the animal host happy. Even so, grass-eating animals have to work hard to help their bacterial friends – they have to chew the grass a lot, and many of them have several stomachs so that they can chew, digest and then regurgitate and chew a bit more. Other animals, such as rabbits, use tricks like eating their own poo.

Humans also have a lot of friendly bacteria living in their guts – around 100 trillion of them! That means you have at least ten times more bacteria in your intestines than cells in your actual body. But we have never acquired grass-eating bacteria.

Biologists of the future might be able to engineer a grass-eating bacteria to survive in our guts, making it possible for us to digest cellulose, but even then we probably wouldn't manage to survive on grass. As a defence mechanism to put off grazing animals like rabbits and sheep, grass has evolved a sort of internal armour plating, in the form of silica. Silica is the same stuff that sand and glass are made of, and it makes the grass very tough – so tough that the teeth of grass-eating animals are continuously ground down. Once their teeth get ground too much they can't eat any more and they starve. This is what happens to small grass-eating animals like voles, and is probably why horses went extinct in North America about 6 million years ago: the type of grass available changed to one with high silica content and the horses starved. They only reappeared in the sixteenth century when Spanish invaders brought some with them.

Why do we dream?

Nobody really knows why we dream; it may be a way to rehearse dealing with difficult problems, or it may be linked to learning.

Dreams are among the great mysteries of the human condition, alongside consciousness and the meaning of life. They can be bizarre and boring, inconsequential and full of meaning. In ancient times, dreams were believed to be messages from the gods or the voyaging of the spirit to other worlds. Only recently have sleep scientists learned enough about dreaming to speculate about whether it has any function, and what that function might be.

So what do we know about dreams? Children under the age of ten don't dream that much, but over the age of ten you can expect to have at least four dreams a night, varying in length from five to thirty-four minutes, although the vast majority of our dreams are not remembered. Dreams mostly happen during a phase of sleep known as rapid eye movement (REM) sleep, during which your brain and body are aroused almost as if you are awake. Although you can also dream in other stages of sleep, REM dreams tend to be the most vivid and memorable.

We also know that although dreams can be about anything, there are some common features that pop up a lot. One of the most common dreams is being chased or followed. Dreams usually feature the dreamer as him or herself, and most of the other people in the dream will be familiar in one way or another. Dreams often have bizarre content, in terms of people behaving oddly or things changing their nature. Dreams tend to be very emotional, and mostly they feature bad emotions, such as fear, anxiety and anger.

Sleep experts have a number of theories to explain why we dream and why dreaming evolved in the first instance. The psychoanalytic theory says that dreams help us to deal with emotional and psychological problems, but if this were true wouldn't it help to remember our dreams?

Random activation theory says that dreams have no meaning or purpose, and are just the by-product of what happens to the brain when it goes to sleep. But if this is true, why have we evolved to have REM sleep, which uses up a lot of energy – almost as much as being awake? By itself, this suggests that there must be a very good reason why we dream.

There is some evidence for this third theory: people who learn new things during the day are likely to have more and longer periods of REM sleep in the night. But there isn't really any evidence that dreams are particularly helpful, or that people improve after dreaming.

Idle time

A third theory, or collection of theories, is that dreaming is a bit like what a computer does when you are not using it. If a computer sits idle for long enough it switches into a kind of housekeeping mode, where it cleans its memory, updates programs and does other 'off-line processing'. According to this school of thought, dreams help your mind to 'clean up' difficult emotions and thoughts, providing psychological healing, and they also help you with learning by filing away new memories and knowledge in the right place.

Another set of theories says that dreams evolved because they helped early humans to perform better during the day, perhaps by letting people rehearse situations and how they would deal with them in a kind of virtual reality simulator. In particular there is a theory called the threat simulation theory, which says that dreams are a safe way to practise how to deal with threatening situations. This would explain why negative emotions are such a common feature of dreams, and why so many dreams are about being chased or being in danger.

Why is the sky blue?

The sky is blue because air scatters blue light but lets other colours pass straight through.

Viewed from space the sun is supposedly a 'peach pinkish' colour, while the sunlight that hits the Earth is white. In scientific terms, colour corresponds to a particular wavelength of light and so white is a mixture of wavelengths. The colours that we see are the result of what happens to sunlight as it passes through the atmosphere, all 5,200 trillion tonnes of it – that's about 25 million tonnes of air for every square mile of the Earth's surface.

If you were to look directly at the sun (which you should never do) you would see yellowish white light. This is what is left of the sunlight that is coming directly to your eyes after the wavelengths of blue have been scattered by hitting air molecules. The red and yellow light is hardly scattered at all, and so comes directly down to Earth in a straight line. The blue light, on the other hand, bounces about all over the sky and some of it eventually reaches your eye. So if you look anywhere other than directly at the sun, you will see blue light.

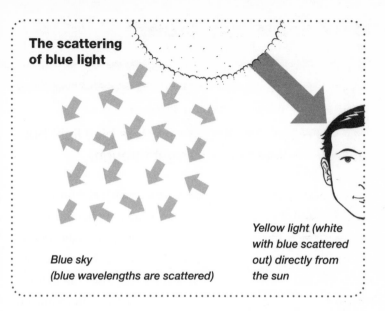

The scattering of blue light

*Blue sky
(blue wavelengths are scattered)*

Yellow light (white with blue scattered out) directly from the sun

As the sunlight – with its mixture of colours/wavelengths – passes through the atmosphere, it hits air molecules and some of it gets scattered. This effect is called Rayleigh scattering, after the British physicist John William Strutt, Lord Rayleigh (1842–1919), who worked out why it happens. The amount of scattering that happens depends on the wavelength of the light and the size of the particles that are doing the scattering. Very tiny particles, like the oxygen and nitrogen molecules that make up 99 per cent of the atmosphere, scatter short wavelengths of light (blue colours) but not long wavelengths (red). The same effect can be seen in smoke made up of very tiny particles, and this is why tobacco smoke appears blue. This colour is named after a different physicist, John Tyndall (1820–93), and is known as Tyndall blue. In fact, it was Tyndall who first explained why the sky is blue, even though Rayleigh gets all the glory.

Rayleigh scattering also indirectly explains why sunsets are red and orange. When the sun is going down over the horizon the sunlight travels through much more air on its way to your eyes than when the sun is overhead. By the time the sunlight reaches your eyes, all the blue has been scattered out, leaving only the orange and red end of the spectrum. This light does get scattered, but not by air molecules – it gets scattered by larger dust particles. Sunsets become particularly vivid red and orange when the sky is full of dust, such as after a volcanic eruption.

On Mars the atmosphere is very thin so there is relatively little Rayleigh scattering. But there is lots of dust in the air, so red light is scattered instead. This is why on Mars the sky appears red.

Why does ice float?

Ice floats because it is less dense than water thanks to the hydrogen bonds between water molecules.

Water is one of the strangest chemicals in the universe, and many of its strangest features turn out to be essential for life as we know it, particularly the almost unique property of water related to freezing. Unlike nearly every other liquid, water expands when it freezes into a solid, by about 9 per cent, and as a result ice is less dense than water.

This is weird because what is supposed to happen when a liquid freezes is that all its atoms and molecules stop zooming around and settle down into a stable, regular pattern. Imagine trying to cram a lot of people into a phone booth. If they were jumping up and down and waving their arms around you wouldn't fit too many of them in the booth – in other words, there would be a low density of people. But if the people were standing still with their arms by their sides you could pack a lot more of them in, and there would be a high density of people in the booth. This is normally what happens with the particles in a liquid and so they are less dense than solids.

Water is different because the H_2O molecules that make it up have unusual properties. The single oxygen and two hydrogen atoms in each molecule share out their electrons unevenly, and this turns the water molecule into a sort of a magnet, with a negatively charged end and two positively charged ends. Each of these ends can attract and stick to opposite charges and this gives the water molecule the ability to form a type of bond called a hydrogen bond. In particular, each water molecule can form hydrogen bonds with up to four other water molecules.

In liquid water the hydrogen bonding actually allows the molecules to get closer together than the particles that make up most other liquids – about 15 per cent closer. As water cools, its molecules have less energy and they slow down and cluster together, as with any other liquid. This continues until water reaches its maximum density at about 4°C (39.2°F), but below this something unusual happens: each water molecule starts to form all four of its possible hydrogen bonds, so that by the time water freezes at 0°C (32°F) the water molecules are locked into a rigid lattice with a much more open structure than the loose arrangements of molecules in liquid water.

Imagine that all the people in your crammed phone booth grab hold of one another by sticking their arms out straight. They are locked into a solid mass, but they end up further away from one another than if they were jiggling about with their arms by their sides. This is basically what happens in ice.

Ice saver

The lower density of ice makes life on Earth possible. If ice didn't float it would sink. A pond would quickly freeze solid in winter as ice formed at the surface and then sank to the bottom, exposing more water at the top to freeze, and so on. Instead the ice layer floats on top, insulating the rest of the pond. Any fish that live in the pond can survive because the bottom of the pond stays liquid.

On a much greater scale this is what happens in winter in the oceans around the poles. The sea freezes into a layer of ice about 2 m (2.19 yd) thick in the Arctic and 3 m (3.28 yd) thick in the Antarctic, and the ice sheets grow until they cover about 12 per cent of the surface of the oceans, extending over an area of some 15 million km² (6 million miles²) of the Arctic Ocean, and 20 million km² (8 million miles²) of the Antarctic.

The buoyancy of ice also has important consequences for shipping, as the *Titanic* learned to her cost. When glaciers feed into the sea huge chunks of ice calve off the ends, forming icebergs that float in the ocean. In the Arctic about 12,000 icebergs calve off glaciers each year, weighing on average around 1.5 million tonnes, poking 80 m (87.49 yd) out of the water and extending more than 350 m (382.76 yd) below the surface. Fortunately relatively few of them reach the Atlantic, and they shrink by up to 90 per cent by the time they do.

Why do we forget?

We forget so that we can remember other things better.

Forgetfulness is irritating and can be damaging and distressing, but it may also be necessary for basic survival.

Explanations for forgetting depend partly on how memory is explained. The most popular model of memory has three stages, and forgetting can take place at all three. In stage one, information floods into your brain via your senses, and the most important information is held briefly in a very short-term zone called the sensory register. If you don't pay any attention to this information it will almost immediately disappear through a process called decay. Decay is what happens when nerve cells stop firing in a certain pattern, and that pattern is lost.

The second stage of memory formation is called short-term memory. If you are trying to remember a phone number, short-term memory is where you hold it. Again, if you don't make an effort or the information is not memorable, it will decay or simply be replaced by the next thing. In both this case and the previous one, you forget something because you never stored it as a memory.

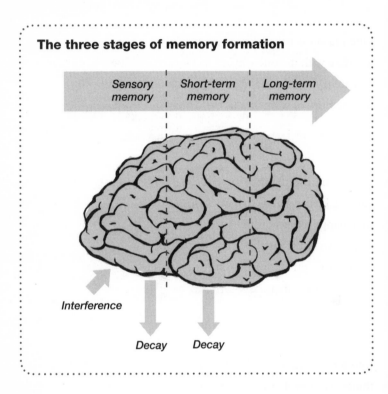

The three stages of memory formation

Sensory memory | Short-term memory | Long-term memory

Interference

Decay   Decay

The final stage of memory formation is long-term memory, in which a memory is laid down in your brain as a more-or-less permanent trace. It seems likely that once a memory is laid down like this it lasts forever unless some physical damage actually destroys the nerve cells involved in the memory. But just because a memory is stored somewhere in your brain, it doesn't mean you can remember it. If you can't get hold of it, it means you've forgotten it, and there are lots of reasons why this could happen.

One reason is called interference. This is where other memories interfere with the one you are trying to remember. For instance, if you are trying to remember what you had for dinner two days ago, the memory of what you ate yesterday might interfere.

Decay and interference explain how forgetting might help us to survive. Decay means that our brains don't store every single thing that happens. According to a November 1983 issue of *Science Digest*, we only remember one out of every 100 things. This helps us to focus on things that are important, such as where to find food, or where that nasty sabre-toothed tiger lives.

Interference makes it hard to remember stuff, so by forgetting things we stop them from interfering with the important stuff. It might be important to remember that meal you ate two years ago, the one with the strawberries you were allergic to, so it would help if you didn't have interference from memories of what you ate last night. Forgetting last night's meal helps you to remember to avoid strawberries. In other words, in order to have a good memory, you have to have a selective memory, and that means forgetting.

Perhaps what is really amazing isn't how much we forget, but how much we can remember. No one knows for sure how much information the human brain can hold, but most estimates agree that your memory can hold around 2.5 petabytes of information (that's 2.5 million gigabytes). A digital video recorder with a capacity of 2.5 petabytes could hold 3 million hours of television programmes. You would have to leave the television running continuously for more than 300 years to use up all that storage.

Why can't we fly?

We can't fly because we are too heavy and too weak.

Birds, bats and insects fly, so why can't we? The simplest answer is that we are too big and heavy – bigger and heavier than any living bird, bat or insect.

Why does size matter? In order to fly you need to overcome the force of gravity. Flying animals generate lift (upwards force) by moving their wings through the air, but to move their wings they need another force called thrust. You need muscles to provide thrust. The bigger and heavier you are, the more thrust you need to produce, and the bigger your muscles have to be.

This is where the problems start. Although bigger muscles do produce more power, the relationship between muscle size and power is not straightforward. A muscle that is twice as big does not produce twice as much power. This is because muscle strength depends on the cross-section of the muscle not its overall size, so that while the mass of the muscle increases to the power of three, strength only increases to the power of two. So to get a muscle that is four times stronger you need one that is eight times bigger.

This is why flight is possible for relatively small animals, but gets increasingly hard as body size increases. Large birds like the albatross and the condor are not strong enough to do much flapping and mostly rely on gliding flight.

According to Rhett Allain, associate professor of physics at Southeastern Louisiana University, a bird as big as a man would need wings at least 7 m (7.66 yd) across, but such big wings themselves would weigh a lot. To flap a pair of 3.5 m (3.83 yd) wings hard enough to lift around 100 kg (220.46 lb) of weight (roughly what a man would weigh if he had 7 m (7.66 yd) of wing) would need muscles many times stronger than the human arm and chest muscles. What's more, because strength does not increase as fast as mass, the muscles in question would be unfeasibly huge, adding yet more weight to the imaginary birdman.

Taking wing

Birds also have a host of other adaptations to help them fly. They keep body weight down with hollow bones. They maintain a very high metabolism with better lungs than mammals, and they use feathers to keep in body heat. They have very large chest muscles, and very large keel bones for the muscles to fix onto.

This wide range of adaptations probably explains how some prehistoric birds, and possibly some prehistoric reptiles, managed to be as big or even bigger than humans. *Quetzalcoatlus*, a pterosaur from the Cretaceous era (about 67 million years ago), had a wingspan of around 12 m (3.12 yd) and may have weighed up to 100 kg (220.46 lb). It was the largest flying animal of all time, but possibly not the heaviest. That accolade belongs to the giant teratorn, *Argentavis*, a vulture-like bird from the Miocene era (6–8 million years ago), with a wingspan of 7.5 m (8.20 yd) and a weight of around 120 kg (264.55 lb).

These monsters could probably barely fly, and must have relied almost exclusively on gliding. The fact that, in terms of weight and wingspan, they match up with our imaginary birdman suggests that maybe he would just about get airborne. Once he had managed this, gliding would be relatively easy. In fact humans with wingsuits can already manage glides of over 16 km (10 miles).

Why is the world getting warmer?

The world is getting warmer because of the greenhouse effect.

A greenhouse works because glass lets light through but traps heat. Sunlight can get through the glass, and when it hits the inside of the greenhouse it heats things up and they radiate that heat outwards. In the open the heat would just escape, but in a greenhouse the glass keeps the heat inside.

Greenhouse gases have the same effect, but on a global level. On a planet like Mars, with a thin atmosphere, sunlight warms the ground but that heat simply escapes out into space. As a result the surface of Mars is a chilly -55°C (-67°F). Earth has a thick blanket of an atmosphere, including greenhouse gases such as carbon dioxide, water vapour and others. These greenhouse gases let through sunlight but trap heat, and as a result the average global temperature is 15°C (59°F); without the greenhouse effect the surface temperature would closer to -18°C (-0.4°F) and life on the surface would be impossible.

> ### *It's getting hot in here*
>
> *The greenhouse effect is essential to the survival of life on Earth, but over the last century or more the Earth has been warming up. The amount and cause of this climate change is very controversial, but the overwhelming majority of scientists agree that the Earth is getting warmer and that industrial pollution by humankind is responsible. This kind of warming is known as anthropogenic, which means 'caused by humans'.*

The two main questions in the debate over global warming are these: is the warming real and is it manmade?

The answer to the first question is that temperatures have increased over the last century, by about 0.75°C (33.35°F) averaged across the globe. Most of this warming has come in the last thirty years, and the ten warmest years on record have come since 1998. Global warming does seem to be real.

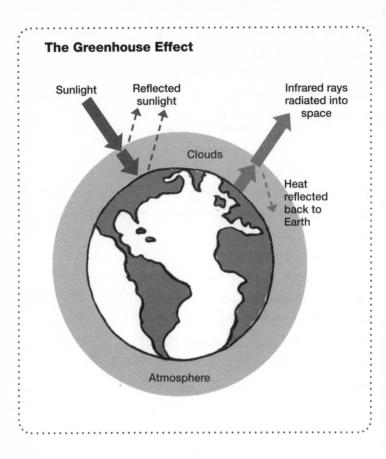

The Greenhouse Effect

Sunlight

Reflected sunlight

Infrared rays radiated into space

Clouds

Heat reflected back to Earth

Atmosphere

The answer to the second question is more complicated. It is impossible to say with 100 per cent certainty because it is not possible to do a controlled experiment comparing our Earth with another Earth where there have been no greenhouse gas emissions from human activity.

We know that human activity currently pumps about 7 billion tonnes of carbon into the atmosphere each year, mostly in the form of carbon dioxide, but we also know that natural carbon emissions, from things like the oceans, volcanoes, forest fires and decaying plants add up to 150 billion tonnes a year.

The difference is that the Earth is used to dealing with these natural emissions, and for many millennia the amount of carbon dioxide in the atmosphere remained constant at around 280 parts per million (ppm). However, human activity such as industrial farming, fossil fuel power stations and cars have pumped extra carbon dioxide into the atmosphere upsetting the delicate natural balance. The carbon dioxide concentration in the atmosphere is now around 387 ppm, an increase of 38 per cent. Two thirds of this increase has happened in the last fifty years. Carbon dioxide levels are now 30 per cent higher than at any time over the last 800,000 years.

The world is getting warmer, and humans are almost certainly to blame. Because the climate takes a long time to respond to changes in greenhouse gas levels, most of the warming is yet to happen. If we continue to produce carbon dioxide at the current rate, the world could warm by up to 6°C (42.8°F) over the next few centuries. According to the British Geological

Society, it will take Earth 100,000 years to recover. Even if we cut our emissions so that carbon dioxide levels don't get over 560 ppm, the Intergovernmental Panel on Climate Change (IPCC) predicts warming of 1.4–5.8°C (34.52–42.44°F) by 2100. The upper limit of this range is more than the difference between the present day and the last Ice Age, while even the lower end of the range would be the biggest temperature change in the entire history of civilization. According to the IPCC, this will probably melt so much ice from Antarctica, Greenland and elsewhere that sea levels will rise anywhere from 9–88 cm (3.54–34.65 in) by 2100.

Why do volcanoes erupt?

Volcanoes erupt because gas-filled molten rock rises to the Earth's surface.

A volcano is an eruption of hot molten rock, solid rock, ash and gas. There are about 1800 active volcanoes in the world: at least twenty of them are erupting right now. When the molten rock comes out of the ground it is called lava, but while it is still in the ground it is called magma. Magma is why volcanoes erupt.

The Earth has a solid crust around the outside, but that crust is pretty thin – only 10–35 km (6.21–21.75 miles) thick in most places. Compared to the whole planet, the crust is very thin. If the Earth were a netball, the crust would be only about half a millimetre thick.

Underneath the crust is a layer of very hot rock that is not quite liquid but not quite solid called the mantle. Sometimes a bubble of magma forms in the mantle. Because it is very hot (up to 1250°C (2282°F) when it comes out of the ground) and full of gas, it is less dense than the rest of the mantle and rises towards the surface. When it gets to the solid crust it forces its way up through cracks.

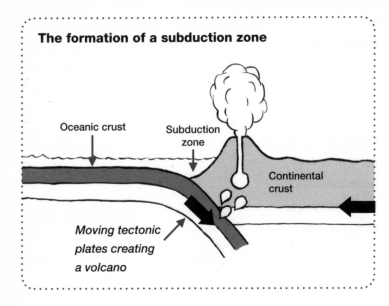

The formation of a subduction zone

Oceanic crust

Subduction zone

Continental crust

Moving tectonic plates creating a volcano

Why does the magma bubble form in the first place? Magma is formed at the boundaries where tectonic plates meet. Tectonic plates are the huge slabs of rock that make up the Earth's crust. They move about relative to one another. Some of them move apart, others ram into each other, forcing the edge of one of the plates to dive down under the other plate. The first type of boundary is called a spreading boundary, the second is called a subduction zone.

Spreading boundaries release pressure in the hot mantle. As the pressure drops the hot rock can turn into liquid and become magma. In subduction zones the subducted crust is plunged deep into the mantle, where extreme heat melts the rock and causes magma to bubble up.

This is why about 94 per cent of known eruptions have been restricted to a few belts of volcanic activity, covering less than 0.6 per cent of the Earth's surface, and these belts correspond to plate boundaries. The best known of these belts is the one surrounding the Pacific plate, which is called the Pacific Ring of Fire; it is home to 90 per cent of the world's volcanoes.

Why are men bigger than women?

Men are bigger than women because our ape ancestors were polygamous.

Polygyny means 'having many wives'. In terms of human evolution, it means that a prehistoric apeman had lots of mates, and he had to fight other apemen to keep them. This supposedly explains why men are 8 per cent taller and 15–20 per cent bigger than women. In fact there are lots of other theories, and some of them lead to amazing conclusions.

Differences in body size are an example of sexual dimorphism. In the world of large mammals, and particularly in the world of our closest relatives, the great apes (chimpanzees, gorillas and orangutans), sexual dimorphism in body size seems to be related to the way that animals choose and keep their mates.

Chimpanzees tend to live in mixed groups of males and females, where different male chimpanzees can mate with different female chimpanzees. Gorillas, on the other hand, have a set-up where one big boss gorilla – a silverback – has a harem of female gorillas that he protects jealously. If another male comes along and tries to poach one of his mates, the silverback tries to fight him off.

This variation in sexual behaviour explains why there isn't a big difference between male and female chimpanzees in terms of size, but there is a big difference between male and female gorillas. Male gorillas have to be bigger so they can fight better and guard their harems, whereas for chimpanzees size is less important than other factors (such as intelligence, perhaps).

Humans and apes shared a common ancestor millions of years ago, so we probably evolved sexual dimorphism according to the same principles. The fossil record shows that today's body size differences between men and women evolved at least 150,000 years ago and probably earlier.

What do differences in size between men and women say about the sexual behaviour of early humans? Body size differences in humans are smaller than in any of the other great apes, so the traditional explanation is that early humans were the most monogamous of the bunch – they were lovers not fighters. But they can't have been completely monogamous, or there wouldn't be any differences at all between men and women.

Not everyone agrees with this traditional explanation. One argument is that although men and women are not that different in size (especially compared to other great apes), there is a huge difference between them in terms of muscles and strength. Men have 60 per cent more muscle than women. Their arm muscles are 80 per cent bigger, which is similar to the difference between male and female gorillas. On average, men's upper body strength is 90 per cent greater than women's, and the average man is stronger than 99.9 per cent of women.

One-woman man?

In fact, we know from looking at societies throughout history and around the world that five out of six of all known societies are polygamous; divorce and remarriage, under some circumstances, are allowed in all human cultures. There is no reason to think things were any different for early humans, so early man probably had to be a bit of a bully when it came to grabbing and keeping a mate.

What does all this mean? According to one explanation, it proves that men evolved to be fighters not lovers. Men are so much stronger than women because they had to evolve to win fights for mates. The strongest men won the fights and got the women, and so genes for male strength were passed on to the next generation.

**The difference in male
and female body sizes**

*Men are roughly 8 per cent taller than
women and 15 to 20 per cent bigger.*

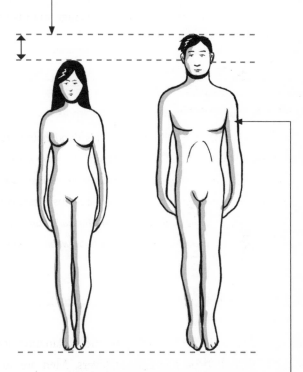

*Men have 60 per cent more muscle
than women. Male arm muscles are
80 per cent bigger.*

There is a problem with this explanation. Genetic studies show that genes for size and strength are passed on equally to both sons and daughters. It seems that there is no way that body size differences between men and women can be passed on to the next generation, since bigger men will have bigger daughters as well as sons.

If this is true, what is the explanation for men being bigger than women? According to a radical theory put forward by Satoshi Kanazawa and Deanna Novak, it's not that men are bigger than women, but that women are smaller than men, and more specifically, that women could grow to be as big as men but they stop growing earlier and so end up smaller.

Their argument is that in a polygamous world, a woman has the best chance of getting a good mate by hitting puberty as early as possible. When a girl hits puberty she stops growing, and this is what restricts the height and size of women. What determines how soon a girl hits puberty? Apparently one of the major factors is how much time she spends with her father growing up. Girls with fathers who are heavily involved in their upbringing tend to delay puberty longer than girls with absent fathers.

The real explanation for why men are bigger than women could be that fathers don't spend as much time raising their daughters as mothers.

Why does iron stick to a magnet?

Iron sticks to a magnet because the magnetic field of the magnet turns the iron into another magnet.

A magnet has two ends or poles called north and south. The north pole of one magnet is attracted to the south pole of another magnet. When an iron nail sticks to a magnet, it is because it has been magnetized, and the south pole of the nail is being attracted by the north pole of the magnet (or vice versa).

When a magnet turns a piece of iron into another magnet, this is called induced magnetism. Iron and other ferrous (iron-like) metals are said to be ferromagnetic because they have properties that mean they can be magnetically induced.

Induced magnetism happens because a piece of iron is made up of thousands of regions, each of which is like a tiny magnet. Normally these tiny magnetic regions, called domains, point in many different directions so their magnetism cancels one another out, but when they are magnetically induced the domains all turn round to point the same way.

A domain is a bunch of neighbouring atoms that all line up together to create a tiny magnet. About 6000 domains fit onto the head of a pin. Each of these domains is made up of around a quadrillion atoms (that's 1,015, or 1,000,000,000,000,000 atoms).

What makes the atoms in a domain magnetic? An atom is made up of a positively charged nucleus surrounded by negatively charged electrons. A magnetic field is produced by an electrical charge in motion so as these electrons spin around, each of them creates its own tiny magnetic field. In most elements different electrons produce magnetic fields in different directions and they cancel one another out, but in ferromagnetic elements, such as iron, cobalt and nickel, the magnetic fields add up and each atom becomes an atomic magnet. Millions of billions of these atomic magnets line up to become domains, and the domains can be induced to line up to turn a whole piece of ferrous metal into a magnet.

Take the magnet away from the iron nail and you take away the magnetic field that is inducing the domains in the nail to line up. The domains go back to pointing in random directions and the nail stops being a magnet.

Some metal alloys, such as steel, keep their magnetism after being induced. This is why you can turn a steel sewing needle into a compass needle by stroking it with a magnet.

The reason that compass needles point north is that the Earth itself is a giant magnet (probably thanks to liquid iron in the Earth's core), but not a very strong one. It is only about a twentieth as strong as a fridge magnet.

Why can't we photosynthesize?

We can't photosynthesize because we don't have chloroplasts, and we wouldn't get enough food out of it to make it worthwhile anyway.

The simple answer to why we can't photosynthesize is that we don't have any of the biological equipment necessary. But there are animals, including at least one vertebrate, which can photosynthesize, so perhaps it's not such an easy question after all.

In plants, photosynthesis takes place in special units inside the cell called plastids. Plastids containing chlorophyll, the green pigment that captures light for photosynthesis (see page 13), are called chloroplasts. Humans can't make plastids – we don't have the genes for it.

It turns out that plants originally didn't have the genes for it either, so 1,600 million years ago they found some bacteria that did have the genes, and invited them to come inside, forever. Plastids were originally photosynthetic bacteria called cyanobacteria, which took up residence inside other cells in a form of relationship known as symbiotic. A symbiotic relationship is one in which both lifeforms benefit from each other. The cyanobacteria benefited from having a safe home with strong cell walls and a constant supply of water; their new hosts benefited from the sugars produced by the cyanobacteria. It was the beginning of a beautiful friendship, and now, 1.5 billion years later, the algae and plants that resulted from this union have conquered the world.

Living on borrowed plastids

If plants are playing host to someone else's chlorophyll, why can't we? There are many species of sea slug that eat algae, suck out their plastids and give them new homes in their skin. They are able to keep the plastids alive for up to ten months before they have to eat some algae to replace them. Sea slugs that do this can live for up to nine months without eating, relying on photosynthesis to feed them instead.

Even more intriguing, biologists have recently discovered that the spotted salamander (*Ambystoma maculatum*) can also play host to photosynthesizing guests inside its own cells. This is an exciting discovery because vertebrates (animals with backbones, such as salamanders and humans) have immune systems that are supposed to stop foreign cells from living inside their bodies. If salamanders can overcome this barrier, maybe humans can too?

With a bit of genetic tinkering or messing about with embryos, we might be able to engineer humans with plastids in their skin cells, so that they can photosynthesize and enjoy a supply of free sugar whenever the sun shines.

Even if we could do this, it probably wouldn't be worth it. Humans don't have that much skin to photosynthesize with, and photosynthesis is not very efficient anyway. According to one analysis, a photosynthesizing human lying naked in the sun at midday for an hour would produce just 15 calories (kcal) of energy, equivalent to about a sixth of an apple. Humans need around 2,400 kcal per day, so in order to survive by photosynthesis alone a green person would have to sunbathe for 150 hours a day!

Why can't we travel faster than light?

We can't travel faster than light because the faster things travel, the heavier they get.

Science fiction wouldn't get very far without warp speed, hyperspace or any of the other ways that spaceships travel faster than light. Unfortunately physics indicates that science fiction can never become science fact because we can never travel faster than light.

Actually it is possible to travel faster than light – in some situations you could outrun a beam of light on a motorbike. This is because light slows down a lot when it travels through other substances. Light slows down to less than half its normal speed when passing through diamond. The slowest speed ever recorded for light is 61.2 kph (38 mph), which was while it was travelling through an exotic form of matter called a Bose-Einstein condensate of rubidium.

What physicists are talking about when they refer to the speed of light is the speed of light in a vacuum, such as in outer space. This speed, referred to as 'c', is 299,792,458 m/s (327,857,019 yd/s).

Why is it impossible to travel faster than the speed of light in a vacuum? The answer is Einstein's famous equation $E=mc^2$ (see page 27). This equation shows how mass (m) and energy (E) are equivalent. It means that the bigger something is, the more energy is locked up inside it.

Crucially, it also means that the more energy something has, the heavier it is. Motion is a type of energy (called kinetic energy), so something that is moving is a tiny bit heavier than something that is at rest. If you throw a baseball at 160.93 kph (100 mph), the ball gets 0.000000000002 g heavier.

This is a tiny amount, but as your speed gets closer to the speed of light, the increase in mass becomes enormous. A spaceship travelling at 90 per cent of the speed of light is twice as massive as the same ship at rest. This means the engines have to work twice as hard to make it go faster. But the faster it goes the more energy it has and the more massive it becomes, and so the more energy you have to put in to speed it up. Meanwhile everything inside the ship is also getting heavier; the watch on your wrist, which used to weigh about 14 g (0.49 oz), would now weigh about 36 tonnes.

A spaceship reaching the speed of light would become infinitely heavy and would need an infinite amount of energy to move. Obviously this is impossible. This is why nothing can go as fast as – let alone faster – than the speed of light.

The top speed attained by human technology was over 240,000 km/h (150,000 mph), by the spacecraft Helios 2

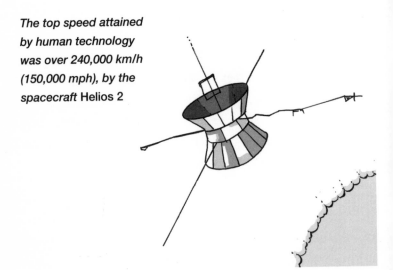

There are some ways around this problem. Some scientists think that particles with no mass such as tachyons could travel faster than light. The speed of light is their slowest possible speed. However, these are purely hypothetical particles and may not exist at all.

It might be possible to get to a distant star faster than a ray of light if you had some way of skipping the space in between. If you could fold or warp space, you could jump from one place to another without crossing the intervening distance. Perhaps it might be possible to go through a wormhole in the fabric of space-time.

Why did Europe colonize the New World?

Europe colonized the New World, and not the other way round, because the Eurasian continent runs from east to west and the Americas run from north to south.

The shape of the continents meant that Europeans had more advanced technology and worse diseases than Native Americans, and that is why Europe colonized the New World instead of the other way round.

In 1492 Christopher Columbus reached the New World and started claiming chunks of it for his Spanish patrons. Over the following centuries European powers and colonists from Europe took control of all of North and South America, and went on to colonize Africa, Australia and Oceania. Most of these places played host to mighty kingdoms and empires of their own at one time or another – indeed the Spanish crushed the Aztecs and Inca, two of the greatest empires in history, in their conquest of Central and South America. So why did Europeans colonize the rest of the world, and not vice versa?

The location of the New World relative to Eurasia

New World

Eurasia

Spain

N
W — E
S

The answer may be linked to the shape of the continents, according to the bold theory of geographic determinism proposed by anthropologist Jared Diamond.

Europe and Asia form a single continent known as Eurasia. Although Eurasia is up to 7,590 km (4,717 miles) across from north–south at its widest point, its north–south axis is for the most part much shorter than its east–west axis, which runs for 10,726 km (6,665 miles) at its widest point.

In the Americas, on the other hand, the north–south axis is much greater than the east–west axis. It is around 17,000 km (10,500 miles) from Prudhoe Bay in Alaska to Tierra del Fuego in Patagonia, but only 4,654 km (2,892 miles) from Point Arena, California to West Quoddy Head, Maine, the widest part of the continental United States.

Why is this important? It means that the main axis of Eurasia crosses many degrees of longitude but stays within a narrow band of latitudes, but the main axis of the Americas crosses a very wide range of latitudes. Climate tends to vary most across latitudes – i.e. along a north–south axis – whereas if you stay at the same latitude you can cross many degrees of longitude without much change in climate.

In other words, Europe, the Near and Middle East, India and China all share broadly similar climate and ecology. Climate and ecology vary massively as you travel from Canada through the Mississippi valley to Central America and Amazonia and on to the Argentinian grasslands. Africa also has a similar north–south axis to the Americas, and a similarly wide variation in climate and ecology.

Horsing around

An important innovation was the domestication of large animals such as horses, camels and cows. Thirteen different species of large animal were domesticated in Eurasia, compared to just one (llamas) in the Americas, and none at all in Africa. The east–west orientation of Eurasia was crucial: it is possible to ride a horse from Spain to China, but a horse will die of disease and overheating if you try to ride it from Algeria to South Africa.

This has had massive implications for the development of human society and health. In Eurasia, technological and cultural innovation were able to spread easily from medieval China to Europe. This is how Europe got hold of world-changing technology such as paper and gunpowder. In the Americas and Africa, it took a very long time for innovations such as metalworking and new crops to get from one empire to another.

As a result, all of the largest empires in world history have been Eurasian. The sixth biggest empire in history, the Ummayad Caliphate of around 750CE covered 13 million km², whereas the largest pre-Columbian American empire, the Inca empire in 1527, covered just 2 million km².

Diseases are also more likely to spread across longitudes but not latitudes, so that people in Eurasia suffered from many epidemic diseases like smallpox and measles (known as herd diseases), but also had the chance to build up immunity to them over a long period.

People in the New World had no immunity, and this is one of the main reasons the Spanish were able to conquer the Aztec and Inca. According to some estimates the population of Mexico fell from 25 million to 6.5 million in just ten years after the Spanish arrived carrying smallpox; the disease killed 75 per cent of the population. In all, European herd diseases probably killed about 65 million Native Americans during European colonization.

Why is the sea salty?

The sea is salty because minerals from the Earth's crust are dissolved by water and washed or blasted into the sea.

A litre of seawater contains about seven teaspoons of salt – that's about 35 g. Only about two and half teaspoons of this is table salt (sodium chloride); in fact there are at least seventy-two chemical elements dissolved in seawater, which almost certainly contains every naturally occurring element on Earth. There are about 50 quadrillion tonnes (50 million billion tonnes) of dissolved solids in the ocean; this is enough salt to cover the land to a depth of 150 m (nearly 164.04 yd).

Rivers and most lakes are nowhere near as salty. Seawater is 220 times saltier than fresh lake water. So why are the oceans so salty while lakes and rivers are not?

In fact, freshwater is a bit salty. When rain falls onto the ground, seeps through the soil and rock and flows over the ground in streams and rivers, it picks up and dissolves minerals. Eventually these are washed out to sea. Rivers and streams flowing from the US alone dump 225 million tonnes of dissolved solids into the sea every year.

Giving vent

Salt also gets into the ocean from hydrothermal vents – fissures in the Earth's surface. So much water blasts out of these vents that if the oceans were drained dry it would only take 10 million years for the vents to refill them. Once the salt gets into the sea it is concentrated by evaporation.

With all this salt flowing into the ocean and so much water evaporating, the real question is not why is the ocean salty, but why isn't it getting saltier? If the oceans lost all their salt overnight, it would only take about 250 million years for the world's rivers to replenish them, yet the rivers have been emptying salt into the oceans for a lot longer than that.

The saltiness of the sea stays the same because the oceans lose salt at pretty much the same rate as they gain it. Salt is lost when it gets laid down as part of new sedimentary rock on the sea bottom, and it is filtered out when seawater seeps down into the crust through cracks in the sea bottom.

Why is the ground solid?

The ground is solid because it is made up of atoms with electrons that cannot be squeezed any closer to the nuclei they orbit.

In 1911 the New Zealand physicist Ernest Rutherford (1871–1937) shot subatomic particles at a piece of gold film. Some of the particles passed right through, but some bounced straight back. Rutherford realized that the atom was not a big solid mass, as had previously been thought. Further research revealed that the atom is like a tiny solar system, with almost all of its mass concentrated into a minute nucleus in the centre, which is orbited by still more minute electrons. Most startlingly of all, there is nothing in between them: 99.999999999999 per cent of the volume of an atom is just empty space!

If you are sitting in a chair or standing on a floor while you read this, think about what is holding you up. Less than one part in 100,000 of that solid surface is actually made of something. The rest is just empty space. Not only that, but all the atoms that you consist of are mostly empty space. There is so much empty space in atoms that if you removed it, you could pack the entire human race into the volume of a sugar cube.

If both you and the floor are mostly empty space, what stops you from falling through the floor?

The reason that solids are solid is that electrons have to obey two laws of quantum physics known as the Pauli exclusion principle and the Heisenberg uncertainty principle. The first law says that two electrons cannot be in the same place – each electron needs its own space. The second principle says that you can know either the location or the speed of an electron, but not both at the same time.

The uncertainty principle means that the more you pin down the location of an electron by confining the space it can occupy, the greater the possible range of speeds it can travel. The electron is like an angry bee in a shrinking box: the more the box shrinks, the more frantically it buzzes about. If you try to squeeze the electron any closer to the nucleus its energy levels go through the roof.

To squeeze a solid by just 1 per cent would raise the energy of the electrons in it by the same amount as heating them by 1000°C (1832°F) This is why you cannot crush a solid into a smaller volume, and hence why solids are solid and you don't fall through the floor.

Why is there no life on Mars?

There is no life on Mars because it is too cold and dry as it has no plate tectonics and so does not benefit from the greenhouse effect.

There may have been life on Mars in the past, when it was warmer and wetter, and there may still be life in Mars, below the surface. But there is almost certainly no life on the surface of Mars.

The immediate reason for this is that it is too cold and dry. The temperature at the surface ranges from -133°C (-207.4°F) to 37°C (98.6°F), but even when it is above 0°C (32°F) water cannot exist as a liquid because the pressure is so low. The Martian atmosphere is 100 times thinner than Earth's, so any liquid water would instantly boil away into gas.

The distance of Mars from the sun is similar to that of Earth – if it had a thicker atmosphere, like ours, the greenhouse effect might keep it warm enough for liquid water and life.

A few billion years ago Mars probably had a thicker atmosphere, but it lost it. Mars is much smaller than Earth and so has weaker gravity, and the atmosphere probably leaked into space. But it might also be because Mars lacks the Earth's plate tectonics.

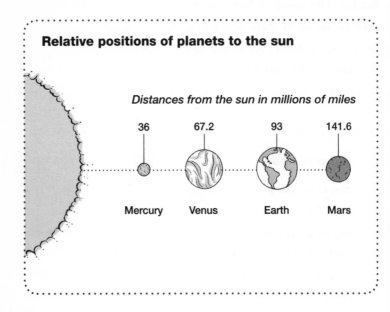

Relative positions of planets to the sun

Distances from the sun in millions of miles

36	67.2	93	141.6
Mercury	Venus	Earth	Mars

On Earth, plate tectonics help to maintain in the atmosphere a constant level of carbon dioxide, the most important greenhouse gas. When water is around, falling as rain or splashing around as seas, carbon dioxide dissolves into the water and then chemically reacts with rocks. This reaction forms rocks like chalk and limestone, and the carbon is locked away. Plate tectonics suck these rocks down into the planet's hot interior, melt them and then blast the carbon back out into the atmosphere through volcanoes. Thanks to plate tectonics, there is a carbon cycle.

On Mars, there is no plate tectonics, and so Mars has no carbon cycle. When it was warmer and wetter, carbon dioxide in the Martian atmosphere was locked away in rocks through similar processes to those found on Earth, but then it was locked away forever. The more carbon dioxide that was removed, the less greenhouse effect there was and the colder it got. The colder it got the more water condensed out of the atmosphere as rain and the more carbon dioxide was removed. It was a runaway reverse greenhouse effect, and the result was a cold, dry, barren planet.

Hot and heavy

Almost exactly the opposite happened on Venus. Venus is almost the same size as Earth, and billions of years ago it too may have had a gentler, wetter climate. But Venus is just a little bit closer to the sun and instead of water falling as rain it stayed in the atmosphere as steam and triggered a runaway greenhouse effect. The heat released masses of carbon dioxide from the rocks and Venus now has a crushingly thick, unbearably hot atmosphere. The surface air pressure is ninety-two times higher than on Earth, so walking on the surface of Venus would be like walking on Earth's ocean bottom a kilometre down, except that the temperature is 477°C (890.6°F). Hot enough to melt lead!

Another reason why tectonic activity is important to life on Earth is that life may well have started around a hydrothermal vent, a sort of water volcano on the bottom of the ocean. The best place to look for life elsewhere in the universe may be around similar vents – for instance, there might be some on Europa, the sixth moon of Jupiter.

Why is the moon the same apparent size as the sun?

The moon is the same apparent size as the sun because even though the sun is 400 times wider than the moon it is also 400 times further away.

Why is there such a remarkable similarity between the size and distance of the moon and the sun? It's a coincidence.

This incredible coincidence is unique among all the planets in the solar system and all 166 of their (known) moons. So far it is unique among all the other planets discovered in the galaxy.

Thanks to this coincidence, Earth witnesses a total eclipse when the moon passes directly in front of the sun, blacking out its disc and leaving only the corona visible.

Total eclipses are rare because the moon is not always the same apparent size as the sun. Both the moon and the earth have elliptical orbits (i.e. not perfectly circular), so usually the moon looks a bit smaller than the sun, leading to what is called an annular eclipse.

Ever since it formed, the moon has been moving further away from the Earth. Today it is moving about 4 cm (1.57 in) further away every year, so if you are fifty years old the moon is about 2 m (2.19 yd) further away now than when you were born.

As the moon moves further away it gets smaller in the sky, so the moon and sun did not match up perfectly in the past and will stop matching up at some point in the future. This means that the sort of total eclipse we witness today, with its spectacular and awe-inspiring corona, could not be seen in the past, and will no longer be possible around 1.5 billion years in the future.

It seems that humans are incredibly lucky to be around at exactly the right time to see total eclipses of the sun. But is it just a coincidence? One possibility put forward is that total eclipses are involved in human evolution, in which case it would not be a coincidence at all.

Why do men go bald?

Men go bald because prehistoric women preferred older men.

There are 193 species of monkeys and apes in the world today, but only one of them is hairless: us. This unusual fact led anthropologist Desmond Morris to come up with his famous phrase 'the naked ape'. In fact humans are not completely naked; we have several patches of hair, most notably on top of our heads. The average person has around 100,000 hairs on his or her head, although blondes have finer, more numerous hairs, while redheads have the thinnest hair.

On average you can expect to drop about sixty-two of these hairs every day, but you can also expect to grow about the same number of new ones. As you get older this cycle of loss and regrowth changes: hairs are lost more quickly while new ones grow more slowly. For a sizeable minority of the population, almost all of them male, the cycle stops completely and lost hairs are never replaced. The naked ape becomes the bald ape.

One in six men go bald, while one in twenty has a receding hairline by the time of his twenty-first birthday. Why is this?

Baldness seems to be genetically determined: you inherit baldness from your father or grandfathers. If we have genes for baldness, we must have them for a reason, otherwise they would have been weeded out by evolution. But what possible evolutionary advantage could come from baldness?

A bald statement

Sociobiologists Frank Muscarella and Michael Cunningham have discovered that women associate bald men with greater social maturity, wisdom and calmness, and this fits with the general stereotype in our society linking baldness and wisdom. Muscarella and Cunningham suggest that some of our apewomen ancestors found these 'bald' qualities more attractive than the immature aggressiveness of younger, hairier apemen, and this explains why baldness not only survived but thrived.

Some other apes and monkeys also go bald, and in chimpanzees there seems to be a relationship between social status and baldness. Bald chimpanzees tend to be respected older males. Perhaps baldness in humans evolved along similar lines.

According to at least one controversial theory, baldness is not really genetic and has nothing to do with evolution. Scientist Yañez Soler suggests that baldness is the result of the wrong type of haircut, which stops hairs from rubbing against each other enough, which in turn damages the hair follicles and sets the unfortunate man on the path to baldness.

The stages of male pattern baldness

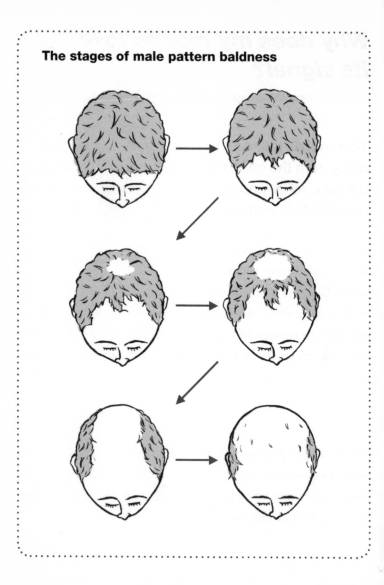

Why does my mobile lose its signal?

Your mobile loses its signal because you move away from the cell tower, buildings and hills get in the way, other people steal your signal and signals can interfere with one another.

A mobile or cell phone is basically a walkie-talkie, except that it doesn't talk directly with other phones; it has to go through a cell tower. Each tower broadcasts radio signals over a relatively small area, or cell – this is why in the US mobile phones are called cell phones. The strength of your signal, and whether or not you lose signal altogether, depends on how well your phone can swap radio signals with a tower.

There are many different things involved in deciding how well your phone and the tower can swap signals: distance, number of people in the area trying to make calls, your network, your phone, the geography of your surroundings and luck.

Each tower has limited power so it can only cover a limited range. It also has a limited number of slots available for use, so if all the slots are being used, your phone will try to connect with a tower that is further away.

This will make a serious dent in your signal strength, because the power at which radio waves are transmitted by a tower drops off as the inverse square of distance. This means that if you have to connect with a tower twice as far away, its signal strength will be four times weaker.

When a signal leaves the tower it has around 20 watts of power, but by the time it reaches your phone it usually has just a millionth or a billionth of a watt of power. So getting bumped onto a more distant tower is bad news.

The weakness of the signal from the tower is also a serious problem if you are inside a building, behind a tree or underwater. Weak signals are easily blocked by brick, concrete, glass, rock and water.

With signals bouncing all over the place and reflecting off things, especially in the city, you could have signals arriving at your phone from all directions at once. These signals are radio waves, and waves can interfere with each other. The trough of one wave can cancel out the peak of another, weakening your signal even more.

Weather can also affect your signal strength. Rain and storms weaken the signals and interfere with your phone's reception. Even the way you hold your phone might affect the efficiency of the aerial, as Apple iPhone users found to their cost during the 'death-grip' furore that greeted the launch of the iPhone 4.

Why do men have nipples?

Men have nipples because women do.

The importance of nipples to women is obvious: they deliver milk to suckling babies – up to 900 ml (1.58 pints) a day! Other mammals manage even more impressive feats of lactation. In 2010 a cow from Wisconsin with the catchy name of Ever-Green-View My 1326-ET produced over 100 litres (175.98 pints) of milk a day for a year! (Cows normally manage around 31 litres a day (54.55 pints).

However, nipples appear to serve no purpose at all in men. Having nipples is part of what defines a mammal (mammals are animals that suckle their young, for which they need nipples), but not all male mammals have nipples. Stallions (male horses) don't have nipples, and neither do male rats or mice.

Men probably have nipples because there isn't a good enough reason for them not to. When embryos are developing in the womb they start off fairly unisex – both male and female embryos begin developing all the same body structures, and these include nipples, which appear around the fourth week of development. By the seventh week, the baby's body begins to assume one gender or another as the sex hormones kick in, causing male babies to develop their sex organs, with the vaginal labia fusing to form the scrotum and the clitoris developing into the penis.

Nipples are not sex-linked body features in the same way as the genitals, so no signals arrive to cause the developing male embryo to lose them. If having nipples were a significant cause of problems for men, or in any way affected their survival and chances of having babies, there would have been pressure via natural selection for them to be lost. But this obviously hasn't been the case, so there has been no reason for men to evolve some special mechanism to lose their nipples during the development of the embryo.

This is a long-winded way of saying that men have nipples because all embryos have them, and all embryos have them because the female ones will need them later.

But this does not explain everything. Male nipples are fully formed and can become functional. During pregnancy and birth a baby is exposed to heavy doses of its mother's hormones, and sometimes high levels of female sex hormones activate the breasts of boy babies so that they are lactating when they are born. This unusual occurrence is called 'witches' milk'. There is also a medical condition called gynecomastia that causes enlargement of male breasts, and which can cause milk to leak from the nipples. Men can also develop breast cancer.

Why spend energy and resources developing fully functional nipples? Wouldn't it be worth evolving away male nipples to protect men from breast cancer? Nobody knows the answers to these questions.

Why does it rain?

It rains because warm air can hold more water than cool air.

When wet air cools down, it can no longer hold as much water, so the water condenses and falls as rain, snow or hail. Temperature drops with increasing altitude, so warm, wet air cools down as it rises. Warm air is less dense than cool air, so warm air will tend to rise on its own. It can also be forced upwards by terrain or by another air mass. Each of these scenarios leads to a different type of rain.

Water can evaporate from a puddle of water. Some of the water molecules in the puddle will be moving about as fast as if they were at boiling point, and some of these will escape from the puddle and get into the air as water vapour. The air can only hold so much of this water vapour; once it reaches its limit the extra water will condense into visible droplets, forming clouds or fog, and some of these droplets will fall as precipitation or rain.

How much vapour the air can hold depends on temperature. Warmer air can hold more, but because it is not as dense as cool air it will tend to rise. The atmosphere cools as you get higher: 6.5°C (43.7°F) for every 1,000 metres (1093.61 yd). As the warm, wet air rises it cools, and above a certain point water vapour condenses into droplets.

Rising damp

Convective rain falls when pockets of warm, wet air rise because they are less dense than the surrounding, cooler air. This occurs when the ground is heated by direct sunlight on a fine day. Warm, wet air can also be forced up as it passes over mountains, causing orographic rain, or it can be forced up when it runs into a mass of cold air. The warm and cold air masses don't mix; instead the warmer one is forced up and over the denser, cold air, creating a weather front, which gives frontal rain. A typical weather front consists of a billion tons of warm air overlaying 750 million tonnes of colder air.

The clouds that form as a result can be enormous: a typical cumulus cloud (the individual fluffy cloud that might be seen floating around on a mostly sunny day) weighs just over 1 million metric tonnes, equivalent to 200,000 elephants. But the amount of water in the cloud can be surprisingly small – our typical cumulus cloud might contain only enough water to fill a bathtub. A storm cloud, on the other hand, can hold 900 tonnes of water.

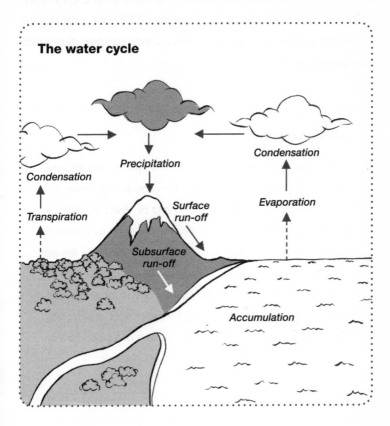

The water cycle

Condensation

Precipitation

Condensation

Transpiration

Surface run-off

Evaporation

Subsurface run-off

Accumulation

Compared to the water in the oceans and ice caps, and even to the water in rivers and lakes, the amount of water in the atmosphere is relatively small: if all the water in the atmosphere fell as rain at once the oceans would only get a couple of centimetres deeper.

But try telling that to the people who live in Mawsynram, India, probably the wettest place in the world, where the average annual rainfall is 1,188 cm (467.72 in), or to the inhabitants of Waialeale in Hawaii, which has an average of 335 rainy days a year. They probably long to visit Arica in Chile, which has one day of rain every six years.

Why are babies and puppies so cute?

Babies and puppies are cute because we are hardwired to respond positively to faces where the eyes are big and the other features small.

If a young woman sees a baby her pupils dilate, and it takes just a seventh of a second for important parts of her brain, involved with the feelings you get when you see something you really like, to light up.

Women aged between nineteen and twenty-six years old might be the group who respond most strongly to cute babies, but everyone is a sucker for puppies, kittens, seal cubs and baby pandas. Even tiny crocodiles or baby skunks seem cute. But what do we mean by cute, and why is it so appealing?

Cuteness is very strongly linked to neoteny, which is a biological term for when creatures keep features associated with very early stages of development such as being newborn or even embryonic. These features are very strong signals that the creature in question is young and defenceless, and needs looking after.

The most obvious cute or neotenous features are ones to do with the face. Young creatures of most species tend to have big heads compared to their bodies, and to have big eyes compared to their heads. In humans, babies have eyes that are almost adult-sized, but their heads are smaller and their bodies relatively tiny. The eyes don't grow much but the rest of the body does, so by the time someone reaches adulthood their eyes are proportionally much smaller.

The cut-off point for baby cuteness is around four and a half years old – children older than this are no longer seen as cute in the same way, and do not trigger the same automatic responses from the adults looking at them. This age marks the point where their features cease to appear neotenous.

For a lucky few, the neoteny and its benefits extend into adulthood. Adults with features such as big eyes and small noses are seen as more good-looking, and men find women with these features more attractive. The effects of neoteny can also cross species barriers, so that humans respond to neoteny in other animals, hence the cuteness of puppies and kittens.

Why should we find cuteness so appealing and attractive? It seems to be hardwired into our brains, as a means for making sure that we bond with and look after very young, helpless babies.

What does this have to do with men finding neotenous women sexy? Men probably evolved to find attractive those features that signal that a woman is young and fertile, and nothing signals youth like a bit of neoteny. Actually neoteny signals too much youth – it doesn't make sense for men to find pre-pubescent girls attractive – but this may be an example of runaway sexual selection. This is where one sex starts evolving certain features that genuinely indicate superior qualities, but once they get going there is a sort of arms race that drives these features to ridiculous lengths. This is why, for instance, male moose have such huge antlers, and possibly why men have evolved such an exaggerated desire for neoteny.

Selected references

AeroSpaceGuide: aerospaceguide.net

Ask a Scientist: askascientist.com

Astronomy magazine: astronomy.com

Astronomy Notes: astronomynotes.com

Environmental Protection Agency: epa.gov

Eric Weisstein's World of Science: scienceworld.wolfram.com

Eureka! The National Children's Museum: eureka.org.uk

Extreme Science: extremescience.com

The Human Touch of Chemistry: humantouchofchemistry.com

Hypertextbook: hypertextbook.com

Mad Sci Network: madsci.org

MATTER: matter.org.uk

Met Office: metoffice.gov.uk

NASA: science.nasa.gov

National Center for Biotechnology Information: ncbi.nlm.nih.gov

National Geographic: nationalgeographic.com

National Institutes of Health, Office of Science Education:
science.education.nih.gov

National Oceanography Centre: pol.ac.uk

National Sleep Research Project:
www.abc.net.au/science/sleep/facts.htm

Nature: nature.com

NewScientist magazine: newscientist.com

Newton, Ask a Scientist, DOE Office of Science: newton.dep.anl.gov

Nine Planets: nineplanets.org

NOAA: www.noaa.gov

Nuclear Energy Institute: nei.org

Office of Scientific & Technical Information: osti.gov

OUP blog: blog.oup.com

The Particle Adventure: particleadventure.org

PBS NOVA: pbs.org/wgbh/nova

Plus magazine: plus.maths.org

Programmed Aging Theory Info: programmed-aging.org

Psychology Today magazine: psychologytoday.com

ScienceDaily: sciencedaily.com

Science magazine of the American Association for the Advancement of Science: sciencemag.org

Scientific American magazine: sciam.com

Scripps Institution of Oceanography, Explorations Now: explorations.ucsd.edu

Society for Popular Astronomy: popastro.com

Stanford Encyclopedia of Philosophy: plato.stanford.edu

The Straight Dope: straightdope.com

The Tech Museum: thetech.org

Universe Today: universetoday.com

University of Illinois Physics Van: van.physics.illinois.edu

US Geological Survey: usgs.gov

WIRED Science: wired.com/wiredscience

Bibliography

American Heritage Science Dictionary, The (Houghton Mifflin Harcourt, 2005)

Banks, William P. (ed.), *Encyclopedia of Consciousness* (Elsevier, 2009)

Chown, Marcus, *Quantum Theory Cannot Hurt You: A Guide to the Universe* (Faber and Faber, 2007)

Comins, Neil, *What If the Earth Had Two Moons?: And Nine Other Thought-Provoking Speculations on the Solar System* (St Martins Press, 2010)

Cullerne, John (ed.), *Penguin Dictionary of Physics*, Fourth Edition (Penguin, 2009)

Davey, Graham, *Encyclopaedic Dictionary of Psychology* (Hodder Education, 2006)

Diamond, Jared, *Guns, Germs and Steel: A Short History of Everybody for the Last 13,000 Years* (Vintage, 2005)

Hartson, William, *The Things That Nobody Knows: 501 Mysteries of Life, the Universe and Everything* (Atlantic Books, 2011)

McFadden, Lucy-Ann, Weissman, Paul R. & Johnson, Torrence V. (eds), *Encyclopedia of the Solar System* (Elsevier, 2007)

Levy, Joel, *Scientific Feuds: From Galileo to the Human Genome* (New Holland Publishers, 2010)

Levy, Joel, *The Doomsday Book: Scenarios for the End of the World* (Vision Paperbacks, 2005)

Matthews, Robert, *Why Don't Spiders Stick to Their Webs?: And Other Everyday Mysteries of Science* (Oneworld Publications, 2007)

McGraw-Hill Concise Encyclopedia of Science and Technology, Sixth Edition (McGraw-Hill Professional, 2009)

Moore, Patrick (ed.), *Philip's Astronomy Encyclopedia* (BCA, 2002)

Pernetta, John, *Philip's Guide to the Oceans* (Philips, 2004)

Pinet, Paul R., *Invitation to Oceanography* (Jones and Bartlett, 2009)

Roeckelein, J. (ed.), *Dictionary of Psychological Theories* (Elsevier, 2006)

Index

(page numbers in italics refer to illustrations)

Also available from Michael O'Mara Books priced at £9.99

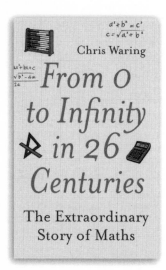

From 0 to Infinity in 26 Centuries:
The Extraordinary Story of Maths
Chris Waring
ISBN: 978-1-84317-873-6

When the Earth Was Flat:
All the Bits of Science We
Got Wrong
Graeme Donald
ISBN: 978-1-84317-868-2